D1601305

Which Way to the Future?

Tor Books by Stanley Schmidt

*Argonaut**
Which Way to the Future?

*forthcoming

Which Way to the Future?

Selected Essays from *Analog*®

Stanley Schmidt

A Tom Doherty Associates Book
New York

WHICH WAY TO THE FUTURE?
Selected Essays from *Analog*®

This book is printed on acid-free paper.

Design by Heidi Eriksen

A Tor Book
Published by Tom Doherty Associates, LLC
175 Fifth Avenue
New York, NY 10010

www.tor.com

ISBN 0-765-30104-0

First Edition: December 2001

Printed in the United States of America

0 9 8 7 6 5 4 3 2 1

To Dennis,
who understands the importance of questioning

Acknowledgments

Analog Science Fiction and Fact has been an integral and important part of my life for most of its duration (part of the time under the name *Astounding Science Fiction*). I am grateful to all of its publishers—Clayton; Street & Smith; Condé Nast; Davis Publications; Bantam Doubleday Dell; and the Dell Magazines group of Penny Publications—for providing a home for it. It has had a long tradition, in its stories and especially in its editorials, of asking questions and looking at familiar ideas from unfamiliar points of view, sometimes even taking a stance that the author didn't really hold for the sake of a good argument; and I'm especially grateful to former editors John W. Campbell and Ben Bova for showing me how it's done. For various contributions to this collection of my own efforts I thank Eleanor Wood, my agent; David G. Hartwell and Moshe Feder, my editors at Tor Books; and Abigail Browning and Scott Lais of Dell Magazines.

The list of people who made some direct or indirect contribution to its content is far too long to include here. Some of them you'll find mentioned in the text, though not always by name; I'd especially like to mention Michael F. Flynn and Charles Goodrich; my parents, Otto and Georgia Schmidt; my brother Dennis; and, as always, my wife, Joyce. Beyond that, the list literally includes almost everybody: all those readers, writers, scientists, artists, politicians, teachers, students, doctors, patients, criminals, engineers, clergymen, naturalists, and people in all walks of life whose discoveries, crimes, insights, mistakes, foibles, and potentials have provided a seemingly endless source of subject matter and inspiration.

Contents

Working to Live, or Living to Work?

Our Environment and Us

Training Our Successors: Myths and Challenges of Education

Introduction by Marvin Minsky

People sometimes ask why I like Science Fiction.

"Why do you read so many books from that tiny niche of literature? When there's so much good writing available, why do you choose to read fiction about mad scientists and their dangerous widgets?"

Almost all of what we call 'literature'—including history, fiction, and politics—is about how people mess up their lives with lust, greed, pride, envy, and treachery—and other ways that we deal with our peers. Of course, those subjects concern us a lot, but are they really so broad and important? Or are all those so-called *humanities* just specialized realms in which members of a particular species tell stories mainly about themselves? That's my view of mainstream writing today: It mostly discusses the same old things. Whereas Science Fiction concerns itself with discussing virtually everything else! As Stanley Schmidt says later in this book in his essay "Nouveaux Clichés," "Remember that human beings are not at the center of the universe, or even the only interesting thing in it."

Critic: *"But the topics that SF so often treats are battles and wars between the stars, or epidemics that strike from afar. Look at Harry Harrison's* Jupiter Plague *and Michael Crichton's* Andromeda Strain, *or the mind-controlling alien beams in Piers Anthony's novel* Macroscope *and in Fred Hoyle's* A for Andromeda. *Surely it makes more sense for us to deal with the problems we already face."*

Of course we have plenty of problems today, but our mainstream writers think mostly about only our recent history, taking short-term views that rarely span even a single millennium. Yet we know that in less than three billion years, the sun will expand to consume the Earth, and that long before that, huge comets will come to eliminate all of humanity. This is just one of the terrible threats that SF tries to prepare us for. As Abraham Lincoln said, "If I had eight hours to chop down a tree, I'd spend six sharpening my axe."

When a child, I read all the SF I could find—including the stories and novels by writers like Jules Verne, H. G. Wells, Aldous Huxley, Hugo Gernsback, Sinclair Lewis, and Philip Wylie. And just when I felt I'd exhausted those teachings, the ideas in *Astounding* began to arrive, as that great editor-writer John W. Campbell acquired his army of powerful thinkers. Month after month new works would appear by such masters as Isaac Asimov, James Blish, Arthur Clarke, Robert Heinlein, Frederik Pohl, Lester del Rey, Theodore Sturgeon, A. E. van Vogt, Jack Williamson, and so many more. These thinkers became like gods to me—along with Galileo, Darwin, and Freud—except for one astonishing difference: These writers also were still alive. At first I worshipped them from afar, but in time it became my privilege to share their friendship and confidence. Here I'll describe a few ways that this helped me develop my own ideas.

In the 1950s I became friends with John Campbell, who eventually decided that the magazine title *Astounding* no longer was appropriate, and he slowly transformed it to *Analog*. Every Thanksgiving holiday, he'd return to his beloved MIT, rent a suite in the Hotel Commander, and hold court with many students and a few professors. I had a ferocious argument with him; he had written an editorial heralding a certain antigravity machine as the answer to low-cost space travel. That machine used a motor to move weights around in a peculiar oscillatory pattern, and the article describing it included two pictures of it on a bathroom scale, showing a lower weight when the motor was on. However, those photographs were clear enough to reveal the name of that bathroom scale, so I purchased one at the local Sears-Roebuck store. Sure enough, the weight would decrease whenever you heartily waved your arm up and down, because there was a mechanical "diode" in the scale's damping mechanism. Claude Shannon and I wrote a letter to Campbell, exposing this flaw in the evidence. Campbell replied with a flaming message about how dogmatic were most establishment scientists.

On another occasion, Campbell introduced me to L. Ron Hubbard, so with some friends I visited his Dianetics institute. Amazingly, there were no 'clears' to be seen; Hubbard explained that they were all too busy for interviews. (He also explained that he himself was not yet able to memorize a page of text in a couple of seconds, as clears were supposedly able to do, because he, as well, had been much too busy to have time to go through his own Therapy.)

At least one of John Campbell's early stories involved building great structures out in space by depositing matter at the intersections of new kinds of energy beams. This could be one thing that influenced me to later become involved with optical physics. See *www.media.mit.edu/people/minsky/papers/ConfocalMemoir. html.*

General literature mostly describes what people most often think about—but it rarely proposes good new ideas about what thinking is or how thinking works. When I first encountered Asimov's ideas, I was entranced by his tales about space and time, but his ideas about robots affected me more. When "Runaround" appeared in 1942, I was already thinking about neural networks and how to make machines that could learn. It was Campbell who proposed those "laws of robotics," and, of course, there was no way to make them consistent, but Isaac treated them mainly as a literary and philosophical device. I think that Asimov's stories were what first taught me about the complexities of commonsense reasoning—which became perhaps my deepest concern throughout the ensuing fifty-odd years. Later in the 1960s, I repeatedly invited Isaac to take a five-minute drive to come over to see how our earliest robots were working, but for years he refused with the excuse that it might hobble his imagination to see robots in such a primitive state. In retrospect, I'm sure he was right, and I still regard him as one of the most sensible philosophers since Bertrand Russell.

Robert A. Heinlein's 1940 novel *Waldo* is what first turned me to thinking about the mechanical aspects of robotics. He described the idea of remote control by putting your arms and hands in mechanical gloves that were connected to similar motorized limbs that were hundreds or thousands of miles away. How it might feel to operate such a system was also described by James Blish in "Bridge" (1952) and by Frank Herbert in *The Dragon in the Sea* (1956); around 1980 I wrote an article about such things in *Omni Magazine*. Telepresence is just now coming of age; it still needs better haptic and tactile sensory feedback so that it can feel like one is actually present at that far away working place. (Most such devices still have only one or two fingers on their hands and do not transmit enough sensation.) I built a variety of telepresence devices in the 1960s and sometimes discussed them with Heinlein himself.

I've also spent a lot of time at meetings for planning space explorations—

some of them with Arthur C. Clarke—because of my feeling that space travel has to become one of humanity's highest priorities. Of course, one motive for doing this is to learn more about the universe. But it's even more urgent to see it as an evolutionary insurance policy, to establish some permanent colonies as far away from Earth as we can get—in case of a cosmic emergency. Indeed, quite a few SF writers have considered the strategy of transmitting our cultures, ideas—and, yes, ourselves—out into the voids of space in the hope that someone will reconstitute us. There is no better depiction of this than the one in Donald Moffit's novels *Second Genesis* and *The Genesis Quest*. Of course, the concept of 'uploading' the contents of human minds has been discussed in many other SF works, for example in Vernor Vinge's novella *True Names*, in stories by Robert Silverberg, Fred Pohl, and Jack Williamson, and in my own *The Turing Option*, coauthored with Harry Harrison. Most people are concerned, to at least some extent, with prolonging their lives. Some day we'll do this by uploading our minds. See *www.media.mit.edu/people/minsky/papers/sciam.inherit.html*.

I still discuss such subjects with friends like David Brin, Frederik Pohl, Gregory Benford, Harlan Ellison, Harry Harrison, Jerry Pournelle, Larry Niven, Poul Anderson, and Vernor Vinge. Whenever you're facing a mystery, it's important to have at least some ideas—and that is where those great SF writers come in. It doesn't matter if those ideas are wrong: At least they may give you a place to start. Besides, if they seem wrong enough, perhaps you can switch to the opposite! It's reported that the physicist Pauli once said, "That theory is utterly useless. Why, it isn't even wrong!" So I still read more SF than anything else except for technical literature and still see those SF writers as the major philosophers of our time—whereas 'mainstream' writers seem "stuck" to me at trying to find new ways to disguise the same old observations about our frustrations and infatuations.

The most productive scientists are those who both get new ideas and then criticize them.

The best SF writers are those who describe how this can be done.

The most productive editors are—well, see for yourself in this wonderful book by Stanley Schmidt—the inheritor of John Campbell's mantle!

Human and Other Natures:
The Search for Intelligence

King of the Hill
[No Matter What]

Over the years I've heard a great many solemn discussions about "What sets Man apart from the lower animals." After considerable reflection, I've decided that one of the most promising candidates for an answer is, "The fierce determination to believe that there is something that sets Man apart from the lower animals, no matter what it takes to maintain that belief."

This observation is not, of course, entirely original. Sue Savage-Rumbaugh and Roger Lewin give an eloquent account of the history of human determination to view our species as something apart from (and above) the rest of nature in their book *Kanzi: The Ape at the Brink of the Human Mind* (Wiley, 1994). I don't have room to tell the whole story here (though I strongly recommend that you read at least the first chapter [after which you'll probably want to read the rest]), but the highlights are simple enough.

In the beginning was a religiously dominated worldview that took it as an article of faith that Man was, in Mark Twain's words, "the Creator's pet." But even the early evolutionists (and some much later ones) explicitly considered humankind to have several characteristics that were unique—not just quantitatively, but qualitatively different from anything possessed by any other animal. These included language, the using and making of tools, consciousness of self, and culture (broadly defined as the transmission of learned behavior from generation to generation).

During the last few decades, all of these supposedly distinct barriers have been challenged by new observations. Jane Goodall took the unprecedented step of actually watching wild chimpanzees going about their business over long periods, instead of jumping to broad conclusions from a few glimpses, and found clear examples of tool use; many others have been seen since then. Gordon Gallup found that chimps clearly did recognize mirror images as reflections of themselves. Numerous investigators including Savage-Rumbaugh herself (working with bonobos [sometimes called "pigmy chimps"]), and others such as Penny Patterson working with gorillas and Lou Herman with dolphins, have found other animals capable of startlingly sophisticated use of language.

Or so *they* claimed. You might think that such discoveries would be greeted with profound excitement by the discoverers' colleagues, but such has not generally been the case. In practice, the commonest responses to the announcement of new discoveries suggesting that "lower" animals did something previously thought to be exclusively the province of man included extraordinary skepticism, often followed by redefining the activity so that man still qualified and other animals didn't.

For example, when some experimenters claimed that chimps and gorillas had learned sizable vocabularies of symbols and combined them in new ways to communicate with their human associates, other researchers said, "Can't be. They're just mimicking or picking up nonverbal cues from the experimenter." Never mind that what the apes were doing would have been unquestioningly accepted as communication if humans did the same thing. These were "animals," therefore they couldn't really be communicating and what they were doing was "mere mimicry" or "the Clever Hans Effect," even if we had to redefine communication to make it so. A fascinatingly ironic twist on this kind of thinking came when Sue Savage-Rumbaugh had occasion to try teaching retarded humans with the same methods she had earlier used with bonobos—and obtained virtually identical results. (See chapter 7 of *Kanzi*.)

I must emphasize that the kind of resistance I'm talking about is not simply the routine skepticism that is one of the cornerstones of all good science. Any experimental results, especially if they appear to contradict reasonable expectations, *should* be scrutinized quite closely and held to rigorous standards of proof. Much of the criticism of animal language and other behavioral studies seems to go far beyond that, assuming a *priori* that certain results are so impossible that any appearance of them *must* be due to something else. That is *not* good science—and it will affect the results of any research done with intelligent beings.

It seems that a great many people, including quite a few working scientists, have deeply ingrained in their thinking the presumption that there is a firm, absolute boundary between humans and "animals." They apparently feel threatened by any suggestion that that boundary is not as firm or absolute as they thought, and instinctively assume a defensive posture against any such threat. "In one way or another," says anthropologist Matt Cartmill, quoted in *Kanzi*, "policing and maintaining that boundary has been a tacit objective of most paleoanthropological model-building since the late 1940s." The basic strategy for that policing seems to be: First try all possible justifications for not admitting anything else into the fraternity of Those Who Can Do Human Things, and if that fails, change the qualifications.

And now we have a new upstart on the playing field: a "lower animal" of our own invention. Science fiction readers have long been familiar with debates over, "Can a machine be intelligent?" Personally, I've always been amazed and fascinated by the number of people who respond, quite positively and so promptly that they can hardly have given the matter much thought, "Obviously not! A machine cannot think; it can only do what it's been programmed to do."

Now, far be it for me to claim that *everybody* who has reached that conclusion has done so without a great deal of high-quality thought, or even that the conclusion is necessarily wrong. It may well be that there is some fundamental reason that human-built machines cannot think in some sense that humans can. All I can say with authority is that I personally have not yet seen such an argument that I found compelling. And I have observed a history of people saying,

"Computers will never be able to do *A*, and that proves that they can't really think;" and then when computers *do* do *A*, the same people say, "Well, computers will never be able to do *B*. . . ."

We used to hear, for example, that a computer would never be able to beat a human Grand Master in chess. After it happened, we would hear mumblings of, "A computer will never be able to write music that sounds like *real* music. . . ." Yet *The New York Times*, as long ago as November 1997, carried a long article about a computer that did exactly that. An audience listened to the same pianist play three unidentified compositions: one by Johann Sebastian Bach, one by a professional musicologist imitating the style of Bach, and one by a computer program called EMI (Experiments in Musical Intelligence). Asked to say which was which, the audience thought the piece by the musicologist was by the computer, and the piece by the computer was by Bach.

It seems a clear musical version of the Turing Test, in which a machine is to be deemed intelligent if it can carry on a conversation which leads a human to think it is human. Yet many of the reactions were not along the lines, "What an impressive program!" but more like, "How appalling!" The musicologist-pianist found it "very disconcerting" that people could be "duped by a computer program." Douglas Hofstadter, the cognitive scientist who ran the composition contest (and wrote the Pulitzer Prize–winning book *Gōdel, Escher, Bach*), found himself "baffled and troubled" by the experiment, thinking it suggested that "Music is much less than I ever thought it was."

But does it? I don't think so—but it may mean that something else is much *more* than he ever thought. Why should that be frightening? Music remains, quite demonstrably, a powerful way to touch and move human beings. Any human being who can create music that does that has made an impressive and commendable accomplishment. Why should not the same be true, and calmly acknowledged to be true, of a machine that can do the same—or a programmer who can make a machine do the same?

True, the computer didn't do it the same way a human would do it. It did not experience life and then create music that evoked in a listener the same feelings that experience evoked in itself. But does that matter, if it evokes feelings that *could* be evoked by experience? I'm inclined to agree with Dr. David Cope, the composer-programmer who created EMI, who observes, "When I am camping in the Sierras there is an incredible beauty I see. But it is unintended by nature. The plants are not trying to express things to me and the mountain is not trying to communicate. But I'm inspired anyway."

The headline on the continuation of the *Times* article seems to miss that point. It says, "Undiscovered Bach? No, a Mere Computer Program Is the Composer." I say that a computer program that can pass itself off as Bach is far from "mere." How much of the credit should go to the program and how much to the programmers is a subject for another debate; the important point for now is that if human listeners judge the program's output comparable in value to that of Bach, then it *is* comparable to Bach—and that neither diminishes Bach's accomplishment nor threatens the humanity of any of us. It's just another route to a similar end. (And those who insist on feeling threatened but console themselves by saying, "Well, it's not very *good* Bach!" are missing another important point, the one I call the "primitive machine" effect: If our machines can do it this well now, they'll be able to do it better later.)

What fascinates me most in all this is not whether either animals or machines can do things that humans have always thought of as exclusively their own. (My own working answer to that is: "Sure! Why not, and so what?") What fascinates me is why so many of us have such a powerful, knee-jerk need to believe that the answer is *no*. I haven't convinced myself that I know the answer, or at least the whole answer, but I can make some partial speculations that may help point someone in a fruitful direction.

First, I think the answers are at least somewhat different for machines and animals—though not as different as they might first seem. For machines, I think, it's pretty obvious that people feel threatened by the idea that something else might someday be able to do their jobs, or even rule their lives. Change is almost always inconvenient, and most people would rather maintain the status quo, even if they're not terribly happy with it, than take on the risks of making a transition to something new—even if what's beyond the transition might be better. Not many humans can see as far as, say, Joan Slonczewski, who has shown us a future in which "sentients" work alongside humans as equals.

With animals, the situation is a little different. Through our entire history we have become accustomed to pushing them around in ways dictated by our own wants and needs, without much regard for theirs. Many of us, consciously or un, would just as soon keep it that way; and admitting that other animals were closer to our equals than we've assumed would make that harder. As Savage-Rumbaugh and Lewin suggest, with direct reference to animal intelligence experiments, "Man's ability to exploit the planet, to take of its resources as he needs, and to usurp entire forests and all living creatures therein, rests upon the unwritten assumption that the chasm between himself and all other creatures is impassable. All of modern man's activities operate from the premise that the planet is his to allot into countries, states, counties, and individual plots, because he, unlike other creatures, has been given the twin gifts of reason and expression. By assuming that other animals lack these gifts entirely, man obviates any need to listen to the wishes of the creatures with which he shares the planet. He can therefore proceed comfortably by his own lights, blind to information that is perceived as nonexistent."

Ultimately, though, that, too, comes back to a fear of fundamental change. It seems to me that many of us, even scientists, feel so threatened by any talk of animal or machine intelligence because admitting such a thing would probably necessitate profound changes in the way we live and relate to everything else.

And what do such views say about our ability to interact with aliens who are more advanced or better equipped? Savage-Rumbaugh and Lewin talk about our attitude toward "dominion over the Earth"; what will happen if and when we meet, here or there, beings who have evolved similar attitudes toward their own home planets? The reactions I've seen so far to animal and machine intelligence studies suggest that we need some serious attitude adjustments before we're ready for that.

We hear a lot about "self-esteem" these days. Many educational systems have taken to viewing it as an intrinsic good, trying to help students to have it whether they're very good at things or not. But maybe it's time to get back to regarding self-esteem as something to be earned, and earned on the basis of your own virtues and accomplishments rather than somebody else's inferiority.

And maybe it's time that we as a species worked on the ability to respect ourselves on some terms other than being better than everybody and everything else. After all, few individual human beings can enjoy that luxury; probably few species can, either, and it's time we started outgrowing the perceived need for it.

Time-Unbinding

Some years ago, Alfred Korzybski coined the term "time-binding" to describe the human tendency to transmit learned information from one generation to later ones. Korzybski was especially interested in such transmission of information by symbols, such as written or spoken language. That, unlike direct demonstration in real time, makes it possible to communicate about objects that are not present, or concepts that are too abstract to point to—or to communicate with people who are not present, or even not yet born.

In response to my editorial "King of the Hill (No Matter What)," in the July/August 1998 *Analog*, reader Hayward Thresher correctly pointed to time-binding as one of the areas of greatest difference between man and other animals. "What distinguishes man from other animals," Mr. Thresher wrote, "is that he remembers his ancestors and tells of their exploits and lineages. It's an oral tradition which vanishes under the development of writing. (The only reason we know that an ancient Egyptian complained the development of hieroglyphics would lead to the deterioration of men's memory is from the statement being put in writing.)"

This observation is interesting on at least two levels. First, it points out the importance of time-binding. Preliterate humans recognized that if they remembered what their ancestors had done, they could build on it. Thus successive generations could make cumulative progress instead of every generation having to repeat the trials and errors by which the previous one had learned, and therefore progress no further. Second, he is probably right that the general level of memorization skills declined when writing became available—but that's not necessarily a bad thing.

Our preliterate ancestors needed bards who could memorize long epics, and poetic forms to make the process easier, because nothing better was available. Writing *is* something better, provided it's done in a sufficiently durable medium. It serves the same purpose as memorizing rote tales, but does the job better.

If an ancient Egyptian complained as Mr. Thresher suggests, we can know about it because it was written down—and it's most unlikely that we would know if it hadn't been, *even if the oral tradition had been kept at the old level of vitality*. Too much has happened in the intervening millennia; nobody could remember it all, much less have time to recite (or listen) to it all. Writing provides a way to create a communal memory bank far larger than any one mind could hold, and a means for anyone sufficiently interested to access those memories. With writing, we could know what Mr. Thresher's Egyptian thought of this newfangled writing even if his entire civilization and all its descendants died in a plague; with only an oral tradition, all of it would be irrevocably lost.

And the loss would be profoundly important—not specifically for the complaining Egyptian, but for the entire body of knowledge and history his people had accumulated. Time-binding is one of the most fundamental requirements for human progress. As Sir Isaac Newton put it, "If I have seen further than other men, it is because I stood on the shoulders of giants." This is not mere modesty (true or false), but a simple statement of important fact. It's highly unlikely that he would have been able to formulate his laws of gravitation and motion without knowledge of the observations of generations of predecessors, or to develop calculus except as an extension of arithmetic and algebra. Much that we take for granted in our civilization is *much* too far removed from the knowledge base of our early ancestors for any one human to be able to make the leap alone. We have what we have only because the brightest members of each generation have seen ways to add something to the knowledge accumulated by the previous generation, which in turn built on the accomplishments of the one before that . . .

And so on—*way* back.

All of which makes it more than a little alarming to see current trends toward time-unbinding, toward cutting ourselves off from our antecedents—sometimes through simple ignorance or carelessness, sometimes through stupid arrogance.

The "ignorance or carelessness" kind of time-unbinding I dealt with at some length in an earlier essay here ("Continuity," August 1991). The problem here is that in the exuberant flowering of the computer age, we have fallen into the trap of using media that are very fragile and very rapidly changing. An accidental electromagnetic pulse can wipe out whole libraries in a fraction of a second; galloping obsolescence makes stored information inaccessible and unusable in a mere handful of years. Ironically, the very tools that have given us the greatest power in history to manipulate information also threaten to make us the most ephemeral civilization in history. We know a lot about what the Egyptians were doing thousands of years ago; if we continue the worst trends of increasing dependence on fragile and rapidly changing storage media, people may know virtually nothing about us in mere decades.

The good news on that front is that there's a growing recognition of the potential problem, and at least some people are making efforts to make media more durable and to facilitate the transfer of data from older computers to newer ones. So that is not my main concern today. At the moment, I'm more concerned about the "stupid arrogance" kind of time-unbinding.

It took me a while to get that way. When I first started seeing the signs of the problem, the phrase "stupid arrogance" didn't even enter my mind. A few years back, I appeared on a panel at a World Science Fiction Convention about "the growing generation gap in science fiction"—a title that turned out to refer to a tendency for younger readers to know little of older work in the field (and older ones to know little of the newer). At the time, it left me relatively lukewarm, because I hadn't yet realized the significance and prevalence of the problem.

Since then I've become increasingly aware of many young readers, and even writers, who are totally unfamiliar with the work of their important predecessors. It's surprisingly common, for example, to find people who consider themselves very knowledgeable about science fiction yet have no idea who Robert A. Heinlein or Clifford D. Simak was. The trend is by no means limited to science fiction. I've also met musicians who knew nothing of Beethoven, Joplin, or Ellington. I've heard of philosophy professors devoting all their attention to very recent philosophers and completely ignoring the classics, such as Plato.

Or, worse than ignoring, sneeringly dismissing them as unimportant and irrelevant. It has become fashionable among some young writers to take *pride* in having never read even their important predecessors, as if ignorance of what's been done before somehow makes their own work "purer" and therefore better. (It was this realization that called the phrase "stupid arrogance" to mind.) It has become fashionable in too many circles to say, "Why should we care what a bunch of old dead guys wrote? They have nothing important to say to us!" There's a fad for merely shrugging off (without looking at it) the work of "dead white males," as if being dead or white or male could automatically mean anyone with those attributes was not worth reading or hearing.

Well, why *should* people now care what their ancestors did? I've already answered that; please see above. Since advancing an art or a science or a civilization depends on each generation building on what all before it have done, you're unlikely to add anything significant unless you understand where you're starting from. You're not going to do the equivalent of inventing calculus unless you already understand the concepts on which calculus builds. And you're not adding anything useful if you spend your life reinventing those concepts, when they're already available in books you disdained to read, and don't go beyond them.

Those who sneer at the idea that there's any reason for them to learn about the classics—or even the previous generation—are just as wrong as those who thought a classical education was *all* you'd ever need. The fundamental importance of time-binding—the reason it must not be allowed to be discarded, either through ignorance or arrogance—is that it enables us to build on the accomplishments of our predecessors instead of continually reinventing the wheel.

People who have no idea what earlier writers did can—and do—kid themselves that the well-worn retreads of ideas in movies and TV shows are fresh and new, that they originated in those shows instead of being recycled without attribution from much older stories. That greatly reduces the chances that such people will ever come up with anything that really *is* fresh and new. And that's just as true in *any* field where people are trying to do new things with no knowledge of the old.

Admittedly it's hard to be familiar with your cultural antecedents, and gets harder all the time, simply because the amount of "old" stuff is constantly growing. Nobody can be familiar with everything new, much less everything old. But it is possible to learn something about the framework and high points, and it's important to do so. If we throw away the effort to time-bind, civilization can easily degenerate into a new kind of stasis, with its members not continually doing the same thing, but repeatedly cycling through the same old things that their abbreviated memories delude themselves are new.

Nature ~~Versus~~ and Nurture

Even in high school, I thought the "nature versus nurture" argument sounded remarkably silly. How could otherwise intelligent and responsible adults seriously argue over which—heredity or environment—determined people's nature and character? It was all too obvious to me, even with my limited knowledge of genetics and experience with life, that *both* played important parts. Even if it was possible to establish that one or the other played a larger role, it would be folly to pretend that either could be ignored when trying to figure out how people became who they are.

I still think so, so I won't dignify the old argument by resurrecting it and taking one side or the other. Unfortunately, I don't have to resurrect it, because others have already done so. For a while I had naïvely thought this naïvely silly either-or argument had gone away, but it's back in full force. People are once more saying the most ridiculous things, not only with straight faces, but with vehemence that sometimes verges on violence—and with carefree disregard for facts or scientific method.

Analogies are never perfect, of course, but consider how a ten-year-old car became what it is. The first determinant is the design; some cars are simply better engineered than others. No less important is the quality of the materials and workmanship that went into converting the paper plans into an actual structure of metal and glass and plastic and rubber. A car made with careless disregard for the details in excellent plans, or using inferior materials, may be no better than one lovingly crafted from a mediocre design.

And that's just coming off the assembly line. The car's condition and reliability ten years later will depend not just on how it was designed and built, but on how it was driven and cared for. One that has constantly been subjected to screeching starts and stops, and never given an oil change or tune-up, has little chance of being in as good a shape as one driven carefully and maintained conscientiously.

Few people would dispute these claims for cars, because people aren't cars and therefore don't take the arguments personally. But just try making carefully reasoned, closely analogous arguments for people!

And there *are* close analogies. Heredity—the genetic blueprint in DNA—corresponds quite closely to the plans from which a car is built. It defines fundamental limits within which an individual can develop, by defining a wide range of bodily characteristics such as body type and color of skin, hair, and eyes. And, as anyone's experience with those characteristics will readily confirm, it *does* vary significantly from individual to individual. A person with the wrong combination of genes *will* suffer from hemophilia; and the last I heard, nobody had found any way to change that unfortunate fact by tinkering with the victim's environment.

Which is not at all the same as saying that Heredity Determines Everything and Environment Counts for Nothing. While the genetic blueprint defines how a new person is supposed to be formed, the execution of many of its instructions can be modified by environmental influences. This can happen either before birth ("in the factory"), or anytime afterward ("driven off the lot"). A fetus may fail to develop properly either because of genetic mutation caused by influences such as radiation or mutagenic chemicals, or because the mother's diet fails to provide materials the fetus needs to carry out the genetic instructions—much as a car may fail to live up to design specifications because somebody misread a dimension or used flimsy plastic where heavy sheet metal was called for. A car "born" one color may later change to a different one by being repainted or rusting; a person, by spending a lot of time in the sun. A car that never gets enough exercise, in the form of lengthy runs at highway speeds, may eventually suffer from carbon buildup in its engine; a person, from plaque buildup in arteries.

Most people, I suspect, can listen to most of what I've said so far without being violently offended—because the characteristics I've used as examples are "purely physical." The fur begins to fly when you raise the question of whether psychological characteristics are similarly subject to both genetic and environmental influences. One reader told me about a movement to get rid of a university president in his area because he had publicly expressed the opinion that intelligence might be determined, at least in part, by heredity. The reader wrote a letter to the editor of his local paper defending the professor's right to free speech "no matter how repulsive that speech may be," but wondered whether the person he was defending was actually promoting a scientifically defensible theory or just using academic freedom as a cover for racism.

Unfortunately, by the time I expressed an interest in seeing more details of the case, the reader no longer had the original articles. So I can document neither details nor disposition. Let us suppose for the sake of argument, though, that there *was* such a case, and consider its merits. On the basis of the little the reader told me, I see neither racism nor cause for revulsion. The professor, as I understand the reports, seemed merely to be trying to restore a modicum of intellectual balance and honesty to a kind of discourse that has lately been distorted grotesquely by the mania for "political correctness."

The idea that heredity influences intelligence is one that long seemed, on the basis of a great deal of experience, clearly true. In all the recent furor, I have heard no refutation of that experience, nor any evidence that intelligence is independent of heredity. And despite the current fashion for hurling the word "racism" at anyone you disagree with, the idea of a connection between heredity and intelligence does not necessarily have anything to do with race, much less racism.

Intelligence is an *individual* characteristic, determined, like any other human characteristic, by a combination of what equipment a person is born with (heredity) and what subsequently happens to modify the condition and functioning of that equipment (environment). There's plenty of evidence that a brain that gets a lot of stimulation and exercise, just like a heart that gets a lot of stimulation and exercise, tends to work better than one that doesn't. But the fact remains that the initial physical characteristics of either brain or heart—the "hardware," so to speak—are determined primarily by the genetic blueprint from which they are built, possibly with some modification by such things as toxins ingested during pregnancy.

The idea that genetic make-up plays no role in determining individual differences in intelligence is tenable if and only if that genetic blueprint, and consequently the hardwiring of the nervous system, is identical for all human beings. This is not true for any other human characteristic I've heard of; in fact, it can be quite dramatically false. Recent studies have found that individuals' senses of taste vary greatly because of fundamental differences in the number and arrangement of their taste buds (and associated nerves). Why should genetic identity be universal, magically and mysteriously, for intelligence and *only* intelligence? And where is the *evidence* that this is true? Until shown a sound theoretical reason, or observational evidence, or both, I must reject the politically convenient *ad hoc* notion that intelligence, alone among all human traits, is exempt from any genetic influence or variation.

Furthermore, humans have long engaged in breeding other species of animals, such as dogs and cats and horses, to achieve a high probability of selected characteristics—including intelligence. Their methods are well established and continue to be used because *they work*. They don't produce vast numbers of identical individuals whose minds work exactly the same way, regardless of training and experience; but they do enable the production of litters with higher-than-average incidence of higher-than-average intelligence, by whatever measure you might use.

How can humans apply those principles to the breeding of other animals, yet deny that they have any relevance to themselves? Are humans, alone among all animals, somehow exempt from the demonstrable correlations observed in so many other species, even though their heredity is based on the same genetic mechanisms? Where is the *evidence* that this is so? Where is a reason why it *should* be—other than that people want it to be? (Actually, despite the current unpopularity of any terminology that gets anywhere near "eugenics," many people *do* apply the principles of breeding for intelligence and related traits to themselves. They might be horrified if you suggested to them that they were doing so, but the fact remains that bright people planning to raise families tend to seek out other bright people to do it with—and then raise more than the average percentage of brighter-than-average children.)

Nothing in the last four paragraphs has even mentioned "race," so why do critics of people like that allegedly outrageous professor cry "racism"? Here's how my *American Heritage Dictionary* defines that word:

1. The belief that race accounts for differences in human character or ability and that a particular race is superior to others.
2. Discrimination or prejudice based on race.

The only possible connection I can see (and the word "farfetched" leaps to mind) is this. Suppose you grant the possibility of individual intelligence being determined partly by heredity, and you recognize the existence of "races"—a slippery term at best, meaning little if anything more than loosely defined subpopulations that constitute more or less separate gene pools. Then it is conceivable that somebody will do a statistical study and find that the average value of some number purporting to measure intelligence is higher or lower in one group than another, just as the average level of skin pigmentation is higher or lower in one group than another.

I am *not* saying that they would, or that they wouldn't. I don't know, and I don't care. It's irrelevant. *Individual* heredity determines individual qualities, which are what individuals should be judged by. Membership in a particular statistical group—or race, if you want to call it that—does *not* in any way determine an individual's character or ability, or imply that that whole group is in any way superior to any other group, or justify discrimination or prejudice based on race.

In other words, it is a simple observational fact, supported by a well-understood and well-tested theoretical basis, that intelligence does depend to some extent on genetics—and that fact has absolutely nothing to do with racism.

Intelligence is only one area in which nature and nurture have been too often lately linked by the inappropriate word *versus*. When *Analog* published Stephen L. Burns's short story "The Wait" (January 1997), about a woman faced with the decision to have her unborn baby's "gay gene" "fixed," we got an interesting collection of comments on whether nature or nurture caused homosexuality. One reader astutely observed that some people have a desperate need for there to be a gay gene, while others have just as desperate a need for there not to be. Not surprisingly, most of the letters we received seemed to fall into one or the other of those categories.

At least one took us to task for publishing the story on the grounds that "a study has shown" that there is no such gene, but ignored numerous other studies that suggest that there is. At least one other letter cited a study of twins that suggested a strong but imperfect correlation between genetics and sexual orientation; the writer chose to interpret that as meaning that even if there is *some* genetic component, it's unimportant. To some of us with no ax to grind one way or the other, the diverse results of studies so far suggest that both nature and nurture play significant roles, but nobody yet really knows their relative importance or how they interact. Burns's story hypothesized one possible outcome of further research and explored the consequences—which is precisely one of the classic functions of science fiction.

To me, the disturbing point about all these discussions is not any of the possible answers, but the fact that people are trying to force the questions to fit the answers they *want* rather than the ones the universe actually provides. Whether you're talking about intelligence, sexual orientation, or whatever hot button is fashionable tomorrow or next week, the chances are very good that *competent* research will find that it's significantly influenced by *both* genetics and environment. Trying to pretend that it's all one or the other because you want it that way, and trying to justify either viewpoint by flaunting the studies you like and dismissing the ones you don't, isn't science. It's anti-science—and science is one of the most powerful tools we have, one that we abuse at our own peril.

But science is about learning what *is*, not finding systematic ways to convince yourself that things are the way you'd like them to be.

On Being Human

Last summer I took part in a discussion about nanotechnology in which one of the participants posed a rather interesting question. The rest of us had been talking about the myriad ways in which a mature nanotechnology might change future life, a subject already familiar to many *Analog* readers from articles and stories such as Chris Peterson and K. Eric Drexler's "Nanotechnology" (Mid-December 1987), Marc Stiegler's "The Gentle Seduction" (April 1989), and Robert A. Freitas, Jr's "The Future of Computers" (March 1996). Those changes include radical transformations of manufacturing and medicine, for example, that might lead to virtual immortality and unprecedented personal wealth, leisure, and capabilities—along with the potential for a variety of correspondingly unprecedented and deadly abuses.

And the gentleman in question asked, "Suppose we manage to avoid all the dangers and get all the 'benefits.' Won't that mean the end of all that it means to be human? What will people *do* when clever nanomachines do all the work of feeding us and keeping us healthy and making whatever we need?"

The question (paraphrased, since I don't have an exact transcript) deserves an attempt at a serious answer. But before you can answer it, you have to be sure you understand the question. What *does* it mean to be human?

Back in 1959, John W. Campbell had an editorial called, "What Do You Mean . . . Human?" In it he considered the question of how to define humanity and tell when you're looking at it. For example, how would you tell an Asimovian robot how to determine whether the First Law ("Thou shalt not harm a human being . . .") applies?

Campbell proposed several "human" characteristics that might be used as "field marks" to recognize a human being, including the capacity for rational thought, being a living animal, bleeding when cut, having human form and size, lack of fur or hair over most of the body, and the capability for speech. He then proposed applying each of these tests to several beings: an idiot, a baby, a humanoid robot, a chimpanzee, and a man with a prosthetic arm. In each case, at least one of the tests was either inconclusive or gave an answer contrary to the one we all "know" is right. Conclusion: "human" is not a simple concept, and it's far from easy to come up with a clear and satisfactory definition.

But that's not exactly the question that was being asked with regard to nanotechnology. That one was more behavioral. Despite the difficulties Campbell pointed out in defining "human being," most of us are pretty sure we know what and who "we" *are*. But what *do* we do that makes us human?

It's probably safe to say that most of you reading this are human. It's almost equally safe to say that you have spent a sizable portion of your life as a member of a family unit including one adult of each gender and one or more of their offspring. Very likely you, or at least *somebody* in your household, works for a living. You probably understand that to mean that you spend maybe 40 hours per week performing some fairly specific kind of service for other people, who then reward you with pieces of paper which you can then trade to still other people for *their* services, including the production of things you need, such as food. When you're not doing that, probably everyone in your household spends at least some time in some form of play. Most of you maintain various kinds of emotional bonds to your relatives, as well as to nonrelatives with whom you've found some special affinity. You may or may not consider yourself to have a special relationship with a Supreme Being.

All of these things, and likely others which I haven't mentioned, are things you do that you probably think of as making you human, as opposed to a guppy or a geranium. But even in the present world, not all human beings do all those things, or do them in the same way, or attach the same kinds of importance to them. If you look back through history, the variations are even larger.

There was a time when "being human" meant a close attachment to a particular small region of land, and you would never have seen much more than that. To a person raised that way, the casual flitting about the globe indulged in by many modern people might have seemed suspect at least, and the notion of trying to leave the planet altogether might have seemed much too alien to fit into his or her ideas of "humanity." For that matter, in most times and places, "his" and "her" kinds of humanity were quite separate and distinct. There are plenty of people alive as I write this who are most uncomfortable with the present efforts to remove or weaken some of those distinctions. There were times (and there still are places) where "being human" meant not attachment to a particular piece of land, but moving constantly with the seasons and the herds. The measure of a man was how many edible animals he could kill, and the measure of a woman was how many children she could bear and raise.

There is a certain irony in the fact that the man who raised this question in the nanotechnology discussion is a television producer and director. I would love to see him try to explain that to his ancestors who were Paleolithic hunter-gatherers—and not just because of difficulties of explaining technology that they would see as purely and simply magic. I suspect those ancestors had quite definite ideas of their own about what it meant to be human. I think they might have been appalled to find that their descendants had no idea how to hunt aurochs, and in fact lived largely if not entirely on food provided by others.

Yes, a nanotechnology revolution, if one happens, will surely *change* our ideas of "what it means to be human." But so did the taming of fire. So did the development of agriculture, and the settling of large groups of people onto farms and into cities. So did the Industrial Revolution. So is the "Information Revolution" doing even now.

Yet we still consider ourselves human. So did that Paleolithic hunter-gatherer. So did the farmers and feudal lords who followed him. Every one of us has skills and values that would be alien to our ancestors and descendants, and lacks others that those ancestors considered fundamental to human nature. Yet we all consider ourselves human. So, I submit, it will be after whatever far-reaching revolutions still lie ahead of us. The woman who is the focus of "The Gentle Seduction" lives through a revolution (or series of revolutions) every bit as profound as the one that prompted the question that prompted this essay. But, far from *losing* her humanity, she uses the new capabilities to *expand* it far beyond anything she could have imagined at the story's beginning.

What will people do if advanced technology frees them from the need to spend most of their waking hours "making a living"? The question is far from new; every time new advances have led to a reduction in working hours, some people have worried about, "What will they do with their time?"

And most have gone ahead and found answers.

Oh, it's true that for some, at present, the answer consists of watching huge quantities of soap operas or ball games on television. It's fashionable to bemoan this fact, and I admit it's not what I'd want to do with my time—but I can't honestly maintain that what I'd choose to do with my time should determine what everybody else does with theirs. People watching soap operas aren't doing a lot to advance Human Progress, but they aren't murdering and pillaging, either. As long as they're earning their keep, how they spend their free time is their business.

Meanwhile, many people with more leisure time find a wide variety of other things to do with it, from volunteer service to travel to artistic endeavors. These may or may not be "superior" in any objective sense to watching soap operas, but they certainly represent extensions of kinds of activity that have characterized "humanness" all along.

Furthermore, not everything that people do when freed from formerly necessary labor is purely recreational. For a convenient counterexample, bear in mind that the principal effect of nanotechnology that led to this question would be to give us an unprecedented degree of control over our material surroundings. The present culture of the United States already *has* an unprecedented degree of control over its material surroundings, not matched in many other cultures in the contemporary world. Yet it also has an uncommonly high incidence of compulsive workaholic behavior! The fact that people no longer have to spend as much time as they used to on some kinds of work does not necessarily mean that they quit doing any kind of work. Consider the effect the computer has had on a great many people's lives. Jobs that used to be terribly time-consuming are now quick and easy. With those disposed of quickly, the people who do them now have more time for leisure activities—or other kinds of work, including kinds they used to *wish* they could do, but which were not feasible without computer aid.

That's the kind of effect major technological revolutions have always had, and will very probably continue to have. People trying to imagine the effects of the Industrial Revolution, or the computer, tended to think in terms of how the new technology would change jobs they were already doing. They could not foresee that the really important effects would be the *elimination* of many of those jobs, and the development of completely new ones that hardly anyone had yet thought of doing at all. We are almost certainly in a similar position in our efforts to foresee the effects of revolutions yet to come.

Perhaps the most characteristically human habit of all is that we are the critters who actively transform the world we live in—and, along with it, ourselves. The essence of human nature is that we *change* human nature, faster than Nature alone can do it, and in ways that suit our own wants and needs. Future revolutions such as nanotechnology, I suspect, will not change that—though they may change nearly everything else.

The Fermi Plague

I think I know the answer. I hope I'm wrong, because what I'm suggesting is *scary*. But it's uncomfortably plausible. . . .

The question is the one commonly known as "the Fermi paradox" and summed up in three words: "Where are they?" Many lines of scientific research suggest that the evolution of life is a natural and common outgrowth of stellar and planetary evolution, and that interstellar communication and travel should be feasible (though not easy). Playing with reasonable guesses for the relevant numbers makes it seem highly likely that we should by now have had some contact with, or at least clear evidence of, at least one other technological civilization from somewhere other than Earth. There is no generally accepted evidence that we have. So where are they?

A great many explanations have been advanced for our lack of evidence of the existence of extraterrestrial civilizations. (A good survey of several of them can be found in two *Analog* articles by David Brin: "Xenology: The New Science of Asking 'Who's Out There?' " [May 1983] and "Just How Dangerous Is the Galaxy?" [July 1985].) Maybe life, for some reason, is harder to originate than we think. Maybe species that *could* be spacefaring decide not to, for one reason or another. Or maybe interstellar empires avoid contact with us because they fear us or don't want to interfere with our development.

And so on.

The trouble with virtually all the proposed solutions, according to many people who've thought about the problem, is that, while each of them can explain why we haven't heard from *some* civilizations, it seems unlikely that any of them would apply to *every* place with the potential for producing a spacefaring civilization. Of course, it's not necessary to have a single, simple explanation for a "phenomenon" that may really be a whole group of phenomena that look similar at some level. It may be that we've seen no evidence of Civilization A for *this* reason, none of Civilization B for *that* reason, Civilization C for a third, and so on, with the net result that we haven't seen any. But some people think that if even *one* civilization became capable of spreading to the stars, it would fill the galaxy in an astronomically short time. So shouldn't we have seen or heard from *somebody*?

Is there some one thing that might be so likely to happen eventually to *any* technologically advanced civilization that it would account for "The Great Silence"?

I may have thought of one, after reading a series of articles back in February about an anthrax terrorism scare. Someone was alleged to be in possession of enough *Bacillus anthracis* to wipe out the population of New York City, with the intention of doing just that, by spraying it in the subways. No, I'm not going to suggest that every civilization gets decimated by anthrax; and according to follow-up stories a few days after the initial announcement, the February threat turned out to be a false alarm. (The "weapons," we were told, were actually a vaccine.) But it got me thinking. . . .

According to at least some microbiologists, the February bioterrorist threat *could* have been real. Anthrax is one of the deadliest and most easily used of all biological weapons. That's why the U.S. armed forces are being vaccinated against it *en masse*. It's fast-acting and almost always fatal. It's cheap and easy to make in large quantities. (Lawrence Korb, a national security expert at the Brookings Institution, called it "a poor man's nuclear weapon.") And it's easy to load into weapons (which may be as simple as insect sprayers or aerosol cans) to spread it in infectious form to everybody in a large region.

And it's only *one* thing with those properties. Other natural pathogens could be used similarly, if not quite as easily or effectively. Both the constant occurrence of natural mutations and the new capability of genetic engineering suggest that still others may be available in the future that have the same properties to an even greater degree.

In other words, for the first time in history, it's now relatively easy for a single individual to unleash a major epidemic. A small conspiracy, involving a mere handful of individuals, could unleash a *really* major epidemic.

In discussions of the Fermi paradox that I've seen, epidemics have not been considered likely causes of extinction of whole species, much less of *all* whole species. Natural plagues develop slowly enough, and usually locally enough, that some individuals will usually survive and develop resistance to them. A population may be decimated, but enough will remain to let it recover.

*Un*natural plagues may be another matter entirely. A combination of technologies, including biological culturing, weapon-building, and rapid global transportation, can make it possible for a very few individuals (or even one) to do things that really *could* wipe out whole populations, possibly even on a planetwide scale.

It would not be a sane thing to do, of course, but that does *not* mean that nobody would ever do it. People commit suicide every day. Some of them are deeply disturbed individuals, more to be pitied than censured, who see suicide as the only way out of their personal problems, and regret the hurt they inflict on those they leave behind. Others go on rampages and gun down crowds of associates or strangers before themselves. If even one of them is a powerful nut who decides to take *everybody* else with him, and knows how, that's all it takes.

And if populations reach into the billions, the chance of one such individual sooner or later arising in any given civilization is disturbingly high. If that happens fast enough, the average life of a technological civilization may be too short for there to be much chance of two of them occurring close enough together in space and time to make contact. Each one may last only as long as it takes to produce one lunatic with too much power at his fingertips.

Any civilization that wants to avoid becoming one of the casualties will have to find an effective answer to the question: How do you prevent any individual from acquiring or abusing that much power? No question is more important; and, like most important questions, it's not an easy one. And it needs to be answered rather early in a species' technological career—say, at about the stage where our species is now.

Actually, of course, it's at least two questions. Preventing individuals from abusing great power once they have it is a different problem from making sure they never get it. It may already be too late for us to do the latter; and in any case, it's hard to keep genies in their bottles, and even harder to stuff them back in once they're out.

Controlling what people *do* with liberated genies is another thorny question. Most of us would like to be very careful *how* it's done even if we admit that it needs to be done. Individual freedom and opportunity are widely (and, I think, rightly) regarded as some of the most important genuine accomplishments of civilization. Even if we collectively decide that large amounts of both must be sacrificed to make sure that no kook wipes us all out, who's going to make sure the people or agency with the power to exert that control don't abuse it? It's a very old question; I believe the ancient Romans knew it as *Quis custodiet ipsos custodes?* ("Who will watch the watchmen?")

The only really long-term solution, it might seem, is one that many people would reject as an impossible dream. We need to become a world of people who all have the intelligence, mutual concern, restraint, and decency to live together without killing each other *even if we have the means to do so*. So I pose the challenge to everyone out there, and in particular all you writers who do those thought-experiments about possible futures that give *Analog* its name: How can we make that happen? If it *is* an impossible dream, what's the best we can do instead? How can we preserve for everyone as much freedom as possible to build a good life as he or she conceives it, without putting all of us at the mercy of any deranged or evil entity who gets too much power in his hands? If it's an *almost* possible dream, how can the many who *can* be trusted protect themselves from the very dangerous few who can't? Can we figure out a solution *without* the kick in the pants of a planetwide close call?

Not easy questions, any of them. But they're questions to which we need the best answers we can find.

And we need them *now*.

The Art of Arguing

Tilting with Straw Men

A*nalog* is an argumentative magazine.

(Did I hear somebody say, "Oh, yeah?")

Yes, we are. It's a game all of us play: writers, editor, readers. As I've explained several times, but find it necessary to repeat periodically, I usually try to stimulate controversy with my editorials—sometimes by advocating or defending viewpoints I don't actually subscribe to. Most of our writers enjoy turning ideas on their heads and looking at them from unorthodox angles. They delight in asking "What if?" about things that conventional wisdom says we're not supposed to question, and exploring possible cultures that work quite differently from ours. Readers enjoy writing to us, or collaring us at conventions, to tell us when they disagree with something we've said.

As I said, it's a game most of us enjoy. If you didn't, you probably wouldn't be reading this book. Most of us do a pretty good job of keeping the right perspective on it, remembering that it *is* largely a game and should be played with a spirit of fun.

However, it's a little more than *just* a game. The ideas we play with are often not mere arbitrary game pieces, but very real concerns about how the world works and where we as an intelligent species are going in it. We're talking about things like what control we or our governments should exercise over the beginning and end of life, about whether and why and how we should try to go to the stars, and what we should do if we meet anyone else out there.

People *should* argue about such things, because they're going to have to make real-world decisions about them. The more ideas have been thrown on the table and tested against each other, the better our chances of finding *good* options. The more people have tried to pick holes in all the suggestions, the better the chance that the bad ones will be weeded out.

But for that very reason, that such concerns are important in the real world, people also tend to have strong feelings about them. They've heard *some* of the ideas, they've done some thinking, and they've come up with Answers that they've become attached to. Unfortunately, that gets in the way of continuing the discussion or considering new possibilities. When people have decided that they already *know* the answers, they tend to lose patience with listening to other people's suggestions. They don't *want* to play the game anymore. They want to take their chips and go home—and send all those annoying competitors home, too.

So they try to take shortcuts to get rid of the opposition, which often show up as weaknesses in their arguments. Editing a magazine like *Analog*, which surely gets one of the most interesting mailbags in the world, I've had a lot of opportunity to watch *how* people argue. I've noticed that there are certain traps which are quite common, and so insidious that even very intelligent people often fall into them.

One of these is my topic of the moment: the tendency to cite a position or characteristic as belonging to someone you disagree with, and then attack or ridicule that position and consequently your opponent. If your opponent actually *has* that position or characteristic and you describe it accurately, your argument is valid and demands a serious answer. What too often happens instead is that the view or behavior attributed to your antagonist is an oversimplification of what he really thinks or does. In that case, your argument doesn't mean much at all. You're not really attacking your opponent, but only a "straw man" that looks vaguely like him.

Examples are regrettably easy to come by. I've often heard avid supporters of U.S.-style capitalism shrug off "environmentalists" as "tree-huggers" and "environmentalism" as "an odd new form of nature worship," or words to that effect. No doubt this makes the industry-boosters feel smugly superior, but it ignores the fact that there are far subtler forms of environmentalism that cannot be dismissed so lightly by anyone who actually bothers to *listen* to them. On the other hand, there *are* sloppy environmentalists who are horrified at the idea that deer populations need to be controlled and who regard the "military-industrial complex" as an evil monolith, ignoring the fact that some industrialists are very concerned about maintaining a livable environment.

Tilting with straw men is not a weakness characteristic of any particular part of a political or philosophical spectrum. You'll find people doing it on at least two sides of virtually *any* controversy!

Not too long ago we published a story containing several military characters with attitudes that were, shall we say, unhealthy—for other people. We got several letters objecting that military people aren't like that at all, but embarked on their military careers solely because they crave Peace and want to be sure we all get it. Now, there may really be some who did; but I don't for a minute believe that they *all* did, any more than I believe that they're all there solely because of a lust for Power. The real ones I've known have been a good deal more complicated than *either* of those "straw soldiers."

I've found myself bothered on several recent occasions by stories in which demonstrators and activists (always opposing the protagonist!) were uniformly portrayed as simpletons who never did a lick of useful work and participated in demonstrations only because they got a kick out of demonstrating *per se*. Maybe there are some fair approximations of those in the real world, too—but far more important and interesting, in both fiction and reality, are the ones who are there because they are actually very concerned about a particular issue, have done some thinking about it, and think action on it is necessary.

Everybody I've ever talked to who wrote opinion pieces has become used to the occasional piece of mail from some reader who doesn't like his opinions and so tells him all about the psychological hang-ups that "obviously" cause him to hold them. Never mind that this reader doesn't know a thing about the writer personally, or that his "psychoanalysis" requires wild leaps to conclusions that are miles from the truth. It's so much more comforting to believe that the writer is messed up in the head than that he could actually have sound reasons for seeing things differently from the reader!

Fiction writers sometimes run into a similar problem. It happens that in my own writing, I seldom use profanity or explicit sex because I seldom need them for what I'm trying to do, and I see no advantage in turning off readers who might otherwise be interested in what I have to say. I *recommend* a similar course to other writers (unless their subject matter *requires* those elements), because nobody benefits from losing readers. However, I also recognize that, like it or not, the generally accepted standards for what subjects writers can deal with and what language they can use to write about them have changed over the last several decades. I respect the right of other writers to have different interests or priorities than I do. If I generally like a story well enough to buy it, I'm not going to tell the writer that he or she can't use this or that word, or include such-and-so kind of character, because it might offend somebody. (Editors quickly learn that virtually *everything* offends *somebody*!)

When that happens, the few offended readers often couch their complaints in such terms as, "Obviously So-and-so can't write without using @$&* and %¢?!—clearly the mark of an impoverished intellect and vocabulary." The Offended Reader may find it comforting to believe that, but the *fact* is often that So-and-so can and frequently does write without using @$&* and %¢?!, and the Offensive Story is using a *wider*, not narrower, range of vocabulary and character than the reader wants it to. Might the Offended Reader do better to consider *why* an intelligent and gifted writer is doing the things that bother him, rather than making unwarranted and incorrect assumptions about his or her abilities and character?

A good deal of straw man tilting is done across generation gaps, with neither side gaining any real advantage. The fiction problem just mentioned is one example of that, since it often involves somebody who grew up with one set of standards being uncomfortable with one that evolved later. Another common example is the venerable phenomenon of parents who grew up with one style of music dismissing a style popular with their children or grandchildren as "worthless" and its performers as "talentless." There's no reason why they should be expected to *like* the new music—that's purely a matter of taste—but it's a big jump from not liking it to calling its practitioners incompetent. As a musician myself, I have known plenty of others who worked primarily in fields I'm not personally fond of, but were quite capable of doing whatever the musical occasion required. They preferred to *use* their skills for things that didn't appeal to me, but that's an utterly different thing from not *having* the skills.

Oversimplifications and exaggerations are appropriate, even necessary, in satire. Caricatures cannot be judged by the same standards as photographs.

But oversimplifications and exaggerations are quite inappropriate, and counterproductive, in serious argument. Pretending your opponent is a straw man may sometimes help you get what you want in politics, but is unlikely to lead you to the solution which is actually best for a problem. There are at least two dangers that anyone tempted to use such tactics would be well advised to keep in mind. One is that if your listeners recognize that you're oversimplifying and misrepresenting your opponents' views, they may reasonably wonder what you're doing with your own.

The other is that while you're sniping at the straw man you'd like to believe your opponent is, your *real* opponent may be sneaking around behind you with something important to say or do. And that, I would hope, should be disturbing to anyone, no matter which side of an argument they're on.

Two•Stage Process

Many of the activities of civilized humans depend on two or more distinct operations being done correctly and in the correct order. If either is done badly, the entire venture fails. Often the two stages of such a process are done by separate people, so neither person or team alone, no matter how skillful, can ensure overall success. *Both* must get it right, or both their efforts are wasted.

For example, a sturdy house can't be built unless a good foundation is first laid, and then a good superstructure is built on top of it. Its various surfaces must first be properly prepared and then properly painted, or their finishes will not provide the esthetic quality and lasting protection that are their reason for being. If either the preparation or the final painting is faulty, it doesn't matter how good the other is.

Telegraphy is a relatively simple method of long-distance communication that can get a message from point A to point B even under adverse conditions. But it's completely dependent on somebody at point A accurately translating the message into code and keying it into the transmitter, and somebody else at point B accurately translating it back into plain text. If the sender is sloppy, the receiver has only a garbled message. If the sender is fast and fluent but the receiver can't "read" well, the message is lost anyway. They *both* have to do their jobs well—which means both have to know what they're doing and concentrate on doing it well—or communication is not achieved.

I chose that last example with malice aforethought, because telegraphy is one method of communication—but the same principle applies to *any* kind of communication. Communication *per se* is one of the best imaginable examples of something which is inherently a two-stage, cooperative process. The two stages are *always* done by two different people (who are occasionally the same person at different times), and both must do their parts conscientiously and correctly. Telegraphy is a particularly neat, clean example where it's easy to see the two stages and the importance of each. But it's just as true of writing a letter to your Aunt Maizie or reading one from her, or of writing or reading a magazine article.

In any such case, one person translates thoughts into words—symbols—which are conveyed to another, who then translates them back into thoughts. If they both do their jobs perfectly, the thoughts the second person gets from the words are the same ones the first put into them. But for that to happen, the two must have a finely tuned, shared understanding of what words mean. The speaker or writer must choose and assemble his words carefully to accurately express his thoughts. The listener or reader must pay careful attention to the words and the way they're put together to determine what thought they were intended to express.*

In other words, real communication happens *if and only if the writer or speaker takes pains to say exactly what he means, and the reader or listener takes equal pains to read or hear exactly what was said.*

Unfortunately, that all too seldom happens. One or the other, or both, botch it. A large percentage of human conflicts are caused by failures of communication. One person is careless about how he says what he means, and another, even listening carefully, comes away with a wrong impression. Or someone chooses his words very carefully, but someone else overreacts emotionally to one, ignores the rest, and winds up attributing completely inappropriate ideas to the

---

*Note to readers with hyperactive "sexism" detectors: To avoid becoming an example of what I'm complaining about, please read the generic "he" as "he or she" throughout this book. I'm explicitly defining it in the old, generic, nonsexist sense to avoid the awkwardness of repeatedly using three words where one will do. Sentences like the one preceding this footnote are cumbersome enough without adding unnecessary complications!

speaker. Either way, communication fails, even though one person—half the "team" tried diligently to make it work.

The problem is very familiar to writers. I've heard the complaint from every essayist or columnist I've ever heard touch on the subject: that no matter how hard they try to say precisely what they mean, some reader will read something completely different into it. Writers and editors, of course, are not always blameless. Not all of them put as much effort as they should into making sure they express their thoughts clearly and unambiguously, and sometimes even those who do try make a mistake. But quite often they do a good job, and still are misread. They organize their thoughts carefully, translate them into words with meticulous care, and still find readers getting completely wrong impressions. Sometimes such readers go even further and launch into fanciful rhapsodies of amateur psychoanalysis, "explaining" how the writer's opinions are obviously the result of this or that psychological hangup. In some cases they may be; but the assumption is not justified unless the reader knows the writer a lot better than he is likely to from an opinion piece or two.

Why do these things happen? Why are some writers unable to say clearly, precisely, and unambiguously what they mean? Why are even more readers unable to read a paragraph and get from it everything the writer intended to put there, and nothing that he didn't?

After observing and thinking about the problem for many years, I've been forced to the conclusion that much of the blame falls squarely on the very institution which should be doing the most to prevent it: the teaching of English. This is not, I hasten to add, a blanket indictment of English teachers and curricula. Some really do further the development of communications skills both active (writing and speaking) and "passive" (reading and listening). But too many do not. Too many, in fact, are aggressively counterproductive from the standpoint of accurate communication.

In my own experience and observation, both direct and indirect, the more "advanced" an English course is, the more likely it is to actively encourage intentional obscurity in writing and wildly speculative sloppiness in interpretation. I was once told by one of my teachers that I had "wonderful insight" into poetry. Since on the whole I liked and in many respects admired this teacher, I didn't have the heart to tell her that what I really had wonderful insight into was what she wanted to read about poetry. It was easy, for me, to look at a poem and claim to find in it pages and pages of the kinds of Symbolism and Hidden Meaning she wanted me to see there. It was quite rare for me to believe that the poet had actually intended it, or that anything resembling a logical case could be made for that belief.

It seemed obvious to me that if we had to spend two weeks trying to figure out What the Author Was Trying to Say in a page of poetry or verse, he hadn't said it very well. That suggestion, however, was neither welcomed nor taken seriously in classroom discussion. The teacher would rather spend the two weeks and encourage the belief that what we were reading into the piece was true, important, profound, and somehow justified. When we were instructed to cite lines from *Oedipus Rex* or a Thomas Hardy novel as evidence that the characters' lives were controlled by Fate, the teacher would not even listen to an attempt to point out the crucial difference between evidence that their lives *were* so controlled and evidence that they *believed* they were so controlled.

In all these cases, in the name of Literary Analysis, we were encouraged and even required not to paraphrase or analyze what the author actually wrote, but rather to look at it and claim it said impressive-sounding things that it simply didn't. This is not careful communication. It is the antithesis of communication. Similarly, creative writing teachers often encourage their students to write in such a way that no clear meaning can be extracted from their words. They make a virtue of obscurity, perhaps because the more obscure a piece of writing is, the more fun the self-styled analysts can have reading nonsense into it, and the harder it is for anybody to refute them.

Again, this is not communication; it is "anticommunication." And in case any scientists or engineers are smugly applauding my harsh words to somebody else, let me point out that English teachers are not alone in this indulgence. *Any* group that can make a case for considering itself a highly educated elite is likely to cultivate obscurity to further separate itself from outsiders. A former academic colleague of mine told of a graduate seminar he'd taken in which the students took turns lecturing and being critiqued by other students and faculty members. A presentation of his was once downgraded because a questionnaire circulated among the listeners showed that fully 70 percent of them understood what he said. If that many of your listeners understand you, a professor explained, you are obviously speaking at too low a level. (Never mind the alternate explanation, that you were speaking at a good level but with exceptional clarity!)

Now, I must acknowledge that some writers *do* deliberately incorporate complex symbolism and similar devices into their work, and that people are not always fully conscious of all their reasons for doing things. However, in general I consider the author's opinion of what he was trying to say at least as reliable as the fabrications of somebody who doesn't even know him, and in most cases far more so. I have no particular objection to those who find such things amusing, trying to guess what hidden meanings the author might have had in mind, or what unconscious motivations might have influenced him. But those who play that game need to be reminded bluntly and frequently that their guesses *are* guesses, and that it's inaccurate, unjust, counterproductive, and just plain wrong to believe that they are more than that. It's conceivable that someone with a death wish could have the thoughts and feelings in Robert Frost's poem "Stopping by Woods on a Snowy Evening." But there are so many other ways a person could be led to those thoughts and feelings that it's ridiculous to claim, as some have claimed, that the poem *says* he has a death wish.

If these fuzzy-thinking shenanigans and their effects were confined to classrooms, they would bother me less. Unfortunately, what goes on in classrooms has a substantial influence on how people function in the "real world." I see around me a world largely full of people who can neither write clearly and accurately nor read clearly and accurately, when communication requires them to do both. I also see classrooms not only failing to develop those skills, but actively discouraging clarity and accuracy in both writing and reading. The connection seems clear.

If we are ever to have a world in which people understand and empathize with each other enough to work together on solving their common problems, they *must* learn to communicate. Since English and other languages exist primarily as tools for communication, classes in them are the obvious places to teach useful communication techniques and discourage harmful ones. This is not to say that there's no place in them for speculation about hidden meanings, symbolism, and unconscious motivations. But to the extent that they do those things, they need to make students aware that they *are* speculating and must not confuse their speculations with Proven Facts. They must explicitly recognize the possibility that what the author meant to say was exactly what he *did* say. And they must teach students to say exactly what they mean when they seriously want to convey an idea to another mind.

In short, I'd like to see less of our English teaching going into amateur psychoanalysis, and more into giving people the skills to both write and read accurately. So as a new school year approaches and English teachers—indeed, *all* teachers—begin thinking about their course plans, I hope they'll ask themselves: Am I part of the problem, or the solution?

Statistics Abuse

"How many manuscripts," asked the young man in the panel audience at a convention, "do you get in a typical month?"

"Maybe four or five hundred," I answered immediately, since I had heard the question many times before.

"And you publish maybe five or six," observed the young man. "So my chances of selling you a story are maybe one in seventy-five or so."

"Not at all," another panelist broke in before I could answer. "It's *not a lottery*! If you know how to write and know the market, your chances are much better than that. If you don't, they're much worse."

My fellow panelist—an experienced and successful writer who sells at least close to everything he writes—made a valid, important, and often overlooked point. The whole exchange is a good illustration of a common failing of our culture: a tendency to place great stock in figures, especially "odds," with little or no understanding of what, if anything, they actually mean.

The young man's statistics do have a *little* connection with reality, of course, though not a particularly *useful* connection. If you try to guess which manuscripts in my slush pile I will buy, without reading them or knowing anything about my selection criteria, the chances really are about one in seventy-five that any one you pull out of the pile will be one that I buy. (The *precise* meaning is a little more complicated than that, but beyond the scope of this diatribe.)

And that's not, of course, anything like the way I decide the *actual* fate of any particular manuscript. I do that by *reading* each manuscript and deciding by careful consideration of its individual characteristics how I think it would fit into a forthcoming issue of *Analog*. If you were sufficiently familiar with my tastes and quirks and what I'm trying to do with the magazine, you could read any manuscript and come up with a much better estimate than "one in seventy-five" of its chances of being bought.

In many cases your individualized estimate would be very close to one: If I think a story is perfect for the magazine I edit, I *will* buy it if I possibly can. (It will seldom be an *absolute* certainty; sometimes I must turn down a story I like very much because I just bought one that's too similar, or because of a conflict with a book publication already scheduled.) In many cases your individualized estimate would be very close to zero, because some stories are so far from our needs that it's virtually inconceivable that I would buy them. Some simply aren't well enough done; others may be just right for some other magazine, but not this one.

Statistics like "one in seventy-five" are a way of making *some* sort of a guess when you don't know enough about the individual members of a group to predict what will happen to them. If you *do* have such knowledge, you'll do far better to use it—assuming that the detailed behavior of individuals is what you're really interested in, and that you have time to look that closely at individuals.

Sometimes, of course, you aren't or you don't. I use "odds" on manuscripts myself, in one highly specialized way. I do some of my manuscript reading on the train to and from my office in the city. For that purpose, I divide submissions into two piles: the "pros" and "slush." The "pro pile" is manuscripts which I have reason to believe are likely to capture my attention enough to require a very close reading, and perhaps a difficult decision as to whether to buy or not to buy. This might be because the author has had a high batting average in the past, or because I've already started reading a particular story and gotten interested enough to want to read it closely. "Slush" is everything else: stories about which I have no basis for estimating their likelihood of being right for *Analog*.

Please note carefully: this does *not* mean that the stories in either pile receive favored or "anti-favored" treatment. Every story gets read. Sometimes a complete unknown from the slush pile knocks my socks off and gets bought; sometimes an old pro sends me something that I can tell very quickly is not at all what I'm looking for. But statistically, I know that it's almost certainly going to take a lot longer to make decisions about all the manuscripts in a ten-centimeter pile of pro scripts than in a ten-centimeter pile of slush. Therefore I only read "pro" scripts on the train—because a slush pile I could comfortably carry might not keep me busy for the limited duration of a ride.

That's one of the most important uses of statistics: to make valid statements about the *collective* behavior of large aggregates of individuals when you don't know much about the individuals as such. It shines in the branch of physics called statistical mechanics, where you can calculate the macroscopic behavior of a gas, which consists of an *enormous* number of molecules darting hither and yon, without knowing anything about the detailed behavior of even one of those molecules.

But doing such calculations, and interpreting the results, requires a kind of rigor that is too often lacking in statistical arguments by laymen and popular news media—a fact long ago recognized in the title of the book *How to Lie with Statistics*. People who really understand statistics define their assumptions, and the significance of results derived from them, in very precise terms. Most ordinary mortals don't—and if you bandy statistical terminology about it in a sloppy fashion, you can "prove" nearly anything. The trouble is that if information is presented in a way that looks numerical, people take it more seriously than it deserves.

I don't intend to give a formal lesson on statistics, but thought you might be amused and have your thoughts suitably provoked by some actual examples. The one that caught my eye some months ago, inspiring me to plant the seed of this essay on my hard disk, was a full-page ad in a local news magazine, from an organization campaigning to prevent the construction of a sludge treatment plant in a particular location. "As promised in our last ad," the headline trumpeted, "a detailed list of the pollutants that will be produced by Plant X." It proceeded to list thirty-one chemicals or classes of chemicals, each with an "Emission Rate" such as "5.47E-03." Then it said, "[The builder] claims that most of these emissions levels are safe—you decide!"

Sounds objective and fair as all git-out, doesn't it? Except that it gives *absolutely no basis* for deciding. In the first place, there are no *units* on any of its numbers, and therefore no way of knowing what, if anything, they actually mean. In the second place, there is no listing of what levels are considered "safe," and by whom and why, and therefore no way of evaluating the listed numbers even if you knew what they actually were! So the ad challenges readers to "decide for themselves," but gives them no information suitable for doing so. All the average reader is going to see is a long, intimidating list of unfamiliar chemical names and numbers in an unfamiliar notation. It won't even occur to him to ask what, if anything, those numbers *mean*. The sheer length of the list, and its frighteningly "scientific" appearance, are quite enough to make him decide, "We've got to keep that thing out of here!"

Which is, of course, exactly what the advertiser was counting on. (In case you're wondering, I opposed the construction of the plant myself, for reasons of my own—but I was not pleased to see somebody on "my side" using such shabby tactics to fight it.)

More recently I saw a review of a book called *What the Odds Are*, which incorporated several examples of "odds" taken from the book. To be perfectly fair, I should mention that I haven't seen the book itself, so I don't know how accurately the review represents what it says. But I do know that the review very explicitly says, "What are the odds you'll be murdered this year? One in twelve thousand." That *may* be true (I haven't checked) if you consider yourself as a random sample of a suitably defined population. But you could certainly define populations in other (and more meaningful) ways, and conclude that your *individual* chance of being murdered was much higher or much lower, depending on where and how you live.

It's similarly true that about one in a hundred income tax returns is audited—but it's also true that you're much more likely to be audited if your return shows certain peculiarities than if it doesn't. So it is, at best, a gross oversimplification to say flatly, "The odds that you'll be audited this year are one in a hundred." Your odds of ending up in jail (for any reason) are much less than "one in two hundred" if you scrupulously obey all laws, and much more if you go around assaulting policemen in front of witnesses. I'm not sure what "the odds of sinking a hole-in-one are 1 in 3,708" is supposed to mean, but I felt reasonably sure that Arnold Palmer's chances were a good deal better, while those of a non-golfer like me would be much worse.

In other words, practically every example of "odds" in the review is essentially meaningless, unless the underlying assumptions are explicitly stated. But it *sounds* "scientific," without actually requiring any thought, so I suspect the book sold well.

My good friend and esteemed colleague Rowland Shew showed me a newspaper clipping about which he commented, "Unusual for a news item, it includes sufficient data to comment on. To wit: the dichotomy between the headline and the lead sentence versus the actual data. The stacked deck nature of the sample . . ." The headline reads, "Greenpeace Warns of Warming Disaster"; the lead sentence, "Many scientists believe that global warming is running out of control and could lead to total ecological collapse, Greenpeace said." But the *data* say that "Many scientists" means 113 who responded to a survey sent to four hundred scientists "including those on the Intergovernmental Panel on Climate Change and others who have written on the issue in respected journals." So not even the sample to whom questionnaires were *sent* was anything like a random sample of scientists who might have informed opinions on the subject, and 72 percent of those who *got* the questionnaire did not see fit to bother returning it. You can't assume that those 72 percent had the same opinions as the 28 percent who did respond.

There's very little reason to consider such a twice-skewed sample representative of "many scientists." I note also that the news article does not give the actual wording of any of the questions the scientists were asked. I find it hard to believe that many self-respecting scientists would take seriously a questionnaire that asked in so many words, "Do you believe that global warming is running out of control and could lead to total ecological collapse?"

Mr. Shew is also fond of pointing out some interesting statistics in regard to Japan, often maligned for its "closed-market" dealings with the U.S. Ironic, since the actual export totals for 1990 showed Japan buying more from the U.S. than any other country except Canada (and more than Mexico and the U.K. combined). Furthermore, while Japan only bought $48.6 billion worth of goods from the U.S. that year, and the U.S. bought $89.6 billion worth of Japanese goods, remember that there were only 123 million Japanese, but 260 million Americans. *Per capita*, Americans only bought $344 worth of Japanese goods, while Japanese bought $395 worth of American. What was that about a closed market?

I'm running out of space, so I can only briefly mention a few of the other items Mr. Shew (who makes much of his living with statistics) has collected. There is, for example, the refutation of the popular truism that half of American marriages end in divorce—obtained by comparing the numbers of marriages and divorces per year. But that's a meaningless comparison, because most of those divorces are breakups of marriages that occurred in *earlier* years, many of which are still hanging in there. A more careful analysis suggests that only one in *eight* marriages eventually falls apart (though even that is meaningful only in terms of specified ground rules).

Sometimes it's not even a question of "odds," but simply printing blatantly wrong numbers that haven't been checked for internal consistency. Probably most of these go undetected by most readers. How many readers would notice, for example, that a news story reporting that "New Jersey generates approximately 148 million tons of tires a year, about one percent of its total waste stream" implies that the average inhabitant (man, woman, or child) tosses out more than eighteen tons of tires a year? As Mr. Shew observes, "Even allowing for corporate truck fleets, this seems excessive." We leave it to you to calculate the implications if it's true, as someone once claimed, that there are three million homeless persons in the U.S. and forty-five of them die every minute.

Just one more item: a *New Republic* article on the spotted owl scare claimed that the original estimates of alarmingly low populations were obtained by counting owls in a small area and multiplying, while more recent surveys spotted (no pun intended) numerous breeding pairs. It's small wonder they were missed if that's really how the original census was done. That's what statisticians call "faulty sampling"—you can't assume that all chunks of woods that look similar are really equivalent. When I heard this report, I was immediately reminded of an analogous set of results I could get for the Serengeti Plain, which is estimated to have about 1.5 million wildebeest. Wildebeest are herd animals, and migratory. Most of them stick fairly close together and move around a rough circle during a year, following the availability of food and water. If I had tried to estimate the total number by counting the population on $1/n$ of the Serengeti and then multiplying by n, I could have gotten any answer from "wildebeest are extinct" to "billions and billions," depending on which acres I chose to examine.

I could go on and on, but it's really not necessary. Now that you see what you're looking for, I'm sure you'll have no trouble finding plenty more examples of "statistics abuse" in your own information sources. And if you complain about some of them, maybe—just maybe—we'll eventually see some improvement.

Statistics are an extremely useful tool for understanding events and making decisions—but only if you understand what they actually mean.

And what they don't.

Guessing the Future:
A Matter of Perspective

Bold and Timid Prophets

31 May 1996

Dear Mama,

Sorry I missed you when I was in town. As you know from the message on your bleeglefritz, I tried to glog, but you'd already left for your cruise. Hope you had a good time.

I'm writing this on the trazzle, using the sippafiz Myra gave me for my birthday. It's really something; when I was in college there was only one tweedler for the whole campus, and it was nowhere near as powerful as this. I haven't decided yet whether to zilp it to you from the kinkup (I have to change trazzles in Minneapolis) or wait and show off how good I can make a letter look with my frensenoodle. Might as well wait, I guess; I'll be back home in San Francisco by dinnertime, and you'll still be away. (When are you going to get fooba, by the way?)

The reason I was in New York was the national conference on gleedical fizzleworp. Stimulating, but this business travel's getting to be too much. I've racked up over 20,000 miles this month alone! Next month I have to go to London. It'll be my first trip on the PFP, so it will just take a few hours, but I wonder if that will make the galjep even worse.

I've been in touch with my old roomie, Roy Gebivv, on the suplet. He finally got his heart transplant and is doing fine. He's planning to open a new museum of snizzlosophy next month.

On the home front, Pete sliced his thumb clean off in his workshop, but they put it back on and it works almost as good as new. Emily's scheduled to be born June 12, and the sambaloon and doodleburp say she's doing fine.

Well, the doohop just told us to shut off all ditterdeeps in preparation for terwilkening, so I'll close for now. It'll be good to have a few days at home. I just hope I have enough energy to pop something in the nitzeneck when I get there!

Love,
Bob

One of the first things an editor learns is that you can't please anybody with everything, or everybody with anything. Another is that people's *reasons* for complaint are often so different that it's hard to believe that they're talking about the same thing.

Two laments that I hear fairly often at *Analog*, for example, are the following:

(1) There's not as much "sense of wonder" in stories as there used to be.

(2) There's too much "magic," such as characters getting too casually and easily from here to there, by means not clearly understandable in terms of presently known science.

If you think about it, those two complaints seem like opposites, on one level—and very closely related, on another. One seems to say that writers aren't being imaginative enough. The other seems to say that they're letting their imaginations run wild, making suspension of disbelief impossible. Yet a major part of the reason for diminished sense of wonder is that readers' daily lives already include so many things that would have seemed like wild imaginings just a few years ago that it's a lot harder now to evoke the "Gosh-wow!" response.

The letter to Mama at the beginning of this essay is a little reminder of that. It's an "antitranslation" of what we would see as a perfectly ordinary letter that might be written on the date shown, as it might look if it fell into the hands of one of the writer's ancestors in, say, 1860. All I've done is to substitute a nonsense word for every word that would *look* like a nonsense word to that ancestor—and, as you can see, there are a good many such. To us, the original letter would simply say:

Sorry I missed you when I was in town. As you know from the message on your answering machine, I tried to phone, but you'd already left for your cruise. Hope you had a good time.

I'm writing this on the plane, using the laptop Myra gave me for my birthday. It's really something; when I was in college there was only one computer for the whole campus, and it was nowhere near as powerful as this. I haven't decided yet whether to fax it to you from the airport (I have to change planes in Minneapolis) or wait and show off how good I can make a letter look with my laser printer. Might as well wait, I guess; I'll be back home in San Francisco by dinnertime, and you'll still be away. (When are you going to get e-mail, by the way?)

The reason I was in New York was the national conference on genetic engineering. Stimulating, but this business travel's getting to be too much. I've racked up over 20,000 miles this month alone! Next month I have to go to London. It'll be my first trip on the SST, so it will just take a few hours, but I wonder if that will make the jet lag even worse.

I've been in touch with my old roomie, Roy Gebivv, on the net. He finally got his heart transplant and is doing fine. He's planning to open a new museum of holography next month.

On the home front, Pete sliced his thumb clean off in his workshop, but they put it back on and it works almost as good as new. Emily's scheduled to be born June 12, and the ultrasound and amniocentesis say she's doing fine.

Well, the pilot just told us to shut off all electronic apparatus in preparation for landing, so I'll close for now. It'll be good to have a few days at home. I just hope I have enough energy to pop something in the microwave when I get there!

To Great-Grandma, it would look more like the first version. Admittedly she would recognize some of the roots and words in the actual letter, but the combinations in which they occur would be alien and incomprehensible. "Answering machine," for example, is made out of familiar pieces, but conveys no more information than "bleeglefritz." And, of course, the weird-looking parts of the letter extend considerably beyond the words I've translated into gibberish. Great-Grandma would be able to tell what several other phrases *seem* to say, but they would seem like such obvious nonsense or impossibilities that she couldn't believe she was interpreting them correctly, or that her great-grandson really meant them. How, for example, could he have traveled over twenty thousand miles in a month? How could anyone have had a heart transplant or a severed thumb reinstalled? How could he know that the expected baby would be an Emily and not an Edgar?

Both readers and writers would do well to keep all this in mind. It *is* still possible, and desirable, to get a sense of wonder in science fiction; but it's no longer possible to do it by having the characters ooh and aah at their experiences. Now you must come up with an idea that is so striking in and of itself that the *reader* will ooh and aah without being told to. You can't expect that to happen as often as it did when the very ideas of space travel and computers were new and unlikely-sounding to most people. But it can still happen occasionally, and it's an impressive and gratifying achievement when it does. Nanotechnology and virtual reality had that power just a few years ago; but already even they are so familiar that stories about them are, for the most part, ringing variations on a theme. Other ideas with comparable potential are undoubtedly waiting to be thought of, but it won't be an everyday occurrence (and, if the truth be known, it never was).

Truly novel ideas are perhaps *most* stunning when they are demonstrably possible. The fantastic always tickles the imagination, but gains an extra dimension when you also *know* it could really happen. The people who complain about too much "magic" in stories would apparently prefer that science fiction restrict itself to that kind of speculation—but that would be a mistake.

To see why, consider again our present as it would have appeared to someone a hundred or so years ago. Suppose, for example, that you'd been a science fiction editor trying to decide which of several stories to buy about what 1995 might be like. You *could* insist on strict extrapolation—stories dealing only with things that you *knew* to be possible on the basis of the science then known. Your magazine might then feature some fine stories about improved railroads and big steamships and the pollution of cities by a growing number of horses. If you were really on the cutting edge, you *might* even have ventured as far out as radiotelegraphs and automobiles *replacing* some of those horses.

But you would *not* have allowed anything involving atomic bombs, color television, heart transplants, microcomputers, or trips to Mars. You *certainly* would not have tolerated anything about relativity, quantum mechanics, plate tectonics, or genetic engineering. With all of those things *verboten*, you couldn't have dealt with such offshoots as satellite communications, electronic banking, CAT scans, or patented organisms.

In other words, your attempts to imagine possible shapes of the future would have missed far more than they got of what's *really* happened. What you *know* to be possible is likely to be very different from what actually happens. *Some* of it will probably happen, but much of it will be crowded out by things you *didn't* know were possible.

That's why a well-balanced science fiction must include *both* extrapolation—the things you can clearly see are possible—and innovation—the things you can't see how to do, but also can't prove *impossible*. Some of those are things that will turn out to be implicit in known science, but not yet suspected because nobody has yet pursued the implications far enough down a particular road. Radio and television as we know them would have been such things in 1895; they're implicit in electromagnetic theory that was pretty well established by then, but nobody knew yet just how far the implications could evolve. Other fictional innovations might depend on completely new kinds of science that have yet to be discovered. Relativity and quantum mechanics would have been like that in 1895; we can only guess what may occupy similar positions in *our* future.

But we should *try* to guess. If there is an imbalance between the two kinds of speculation in today's science fiction, I'd say that innovation is the part being neglected. There are quite a few writers doing an admirable job of exploring the things we can *prove* our descendants could do. I'd like to see more writers trying just as hard to imagine the *new* kinds of science that might help shape our future—not fantasy, where "anything goes," but new phenomena with new rules that *expand* the possibilities we already know.

Because if we don't do that, our imagined futures are going to miss an awful lot of the bleeglefritzes and fizzleworps.

Defenders of the Faiths

As befits its name, *Analog Science Fiction and Fact* publishes, in a typical issue, at least one article of "science fact." But what does "science fact" mean?

In the strictest possible sense, and the one which some readers would clearly like us to use, it can be interpreted as "an unquestioned and unquestionably true statement about the nature of the universe." Unfortunately, there is little, if any, that we actually know with that degree of certainty. Even if there were, many other readers (and I) would not want *Analog* to define the subject matter of its fact articles so narrowly.

In general, we're not interested in reviews of material so well established that it can be found in textbooks. Such stuff you might as well get *from* textbooks. As a magazine whose main focus is what the future might be like and how we can make it better, we are more interested in subjects on the frontiers of research. With only a dozen fact articles per year, we want to concentrate on things which readers are not likely to have read much about elsewhere and which seem likely to have far-reaching implications for the future.

That, of course, covers a broad range. Some of our articles are descriptions of new technologies or areas of scientific inquiry which are generally recognized as major developments within their fields. Some are about the processes of inquiry and innovation *per se*, or about their legal or social ramifications. And some are about "fringe" work: experiments or theories that would have considerable impact if substantiated, but are currently controversial, sometimes even to the point of being rejected and sneered at by most workers in their fields.

Those bring us lots of letters. Many of those letters fall into one of two extreme camps, both of which miss the point of what the authors and the magazine are trying to do with such articles. So an occasional reminder may be in order. Just what *is* the point of such articles as Jeffery D. Kooistra's "Paradigm Shifty Things" (June 1997) and Dr. Eugene F. Mallove's "Cold Fusion: The 'Miracle' Is No Mistake" (July/August 1997)? Why do we publish them? What purpose do they serve?

To refresh your memory, or in case you missed them, Kooistra's article surveyed heretical work in fields (no pun intended) as diverse as gravitation, electromagnetism, and Egyptology. Mallove's concentrated on the specific area commonly (though perhaps not accurately) known as "cold fusion." Both cited references readers could use to go back to the original work and evaluate it for themselves. Virtually all of those references were by people with the kind of credentials commonly accepted as strong evidence of professional competence, and in Mallove's case there were *dozens* of them.

Yet the mail followed the usual, predictable pattern (and precious little of it showed any evidence that the writers had even looked at the sources cited).

One group wrote to applaud the author for exposing how narrow-minded "The Scientific Establishment" is, or even for proving TSE "wrong." Members of this group often compare TSE to a religious priesthood, and sneer at "orthodox" scientists as "defenders of the faith."

The other "extreme" group, of course, is a subset of those "defenders of the faith" who actually do attack the "heretics" in a largely reflexive way, pointing out that generally accepted theories got that way through a long, hard process of repeatable experiments and peer review—and then dismissing out of hand the possibility that somebody might have found something not adequately covered by those theories. Such correspondents sometimes go on to attack *Analog* for being "antiscientific" for daring to publish such hogwash—a view I find ironically amusing since I am myself a scientist. (I left that line of work only because editing *Analog* sounded like too much fun to pass up; I have never ruled out the possibility of someday returning to it, and the advancement of scientific understanding has always remained one of my primary concerns.)

Both groups act like defenders of different faiths: in one case the faith that science already understands things so well that uncomfortable data must be wrong and can safely be ignored; in the other, the faith that science is "just another priesthood" and someday the downtrodden underdogs of independent research will prove the scientific priests wrong.

In fact, *both* groups are wrong in important ways. Both could use reminders of important facts about how science actually develops—and how they can best contribute to that development.

Science is *not* "just another priesthood," or just a matter of faith. It rests on a foundation of accumulated data and critical analysis, and generally accepted scientific theories usually do have a lot of validity in that they describe the data to date more completely and accurately than other theories that have been proposed. The authors of most of our "heretical" articles understand that (those of the two articles I used as examples are scientists themselves), and critics who attack them by pointing it out are simply poking at straw men and undermining their own credibility. Our authors seldom, if ever, deny that science works that way, or claim to prove that established theories are "all wrong."

They do, however, remind us that new observations sometimes turn up that are not adequately explained or predicted by the currently accepted theories, and that the theories must be revised, expanded, or replaced to account for them. That, too, is an important part of how science advances. An old theory seldom has to be scrapped entirely; if it was solidly based on one range of experience, the new theory must give the same predictions as the old in the range of experience where the old theory worked. Newtonian physics, for example, remains a useful model for slow-moving, macroscopic objects, even though those cases are equally well (but less conveniently) described by relativity and quantum mechanics. But observations on very small and/or fast-moving objects *require* the additional subtleties of relativity and quantum mechanics, and those theories are based on a worldview radically different from Newtonian mechanics. People who forget that new data sometimes require new theories are just as wrong as those who forget that science is based on evidence, not blind faith.

The more perceptive of those who write to attack our "heretics" sometimes point out that, with the sheer volume of information being generated these days, few scientists can afford to spend much time looking closely at work which appears unlikely to withstand scrutiny. This is quite true, and a real problem; yet you can't really be *sure* what will withstand scrutiny unless and until you scrutinize it. I addressed this dilemma years ago, in an editorial called "Advice for 'Crackpots' " (which follows), wherein I explained the situation from a professional scientist's viewpoint and made concrete suggestions as to how a proponent of an unorthodox but meritorious view might enhance his or her chances of getting a serious reading.

On the other hand, the fact remains that science is done by human beings, and for some of them the dismissal of the unorthodox as "not worth my time" has become so automatic that it's practically impossible to get them to seriously consider *any* unorthodox work. This, too, is understandable; if you've seen enough letters from people who claim they've disproved relativity, but soon make painfully clear that they don't even understand what relativity *says*, it becomes harder and harder to approach yet another with an open mind. Yet it will occasionally be crucial to the advancement of science to do exactly that. Obviously *everybody* can't listen attentively and mull over *every* unorthodox suggestion. But the whole point of scientific publication is to get new data and ideas out so that, somewhere among the many potential readers, they may find a few who latch onto them and check them out.

Several people who wrote about our latest "heretical" articles objected to what they saw as stridency of tone (which can indeed be counterproductive), and reminded us that if an odd new observation is real enough, it will eventually assert itself so insistently that it can no longer be ignored and theories will *have* to revised to accommodate it. *But that can't happen* if a particular kind of new observation is systematically blacklisted so that it *isn't* repeatedly called to the appropriate people's attention. That *can* happen, and there is at least some evidence that it has happened in the case of "cold fusion."

According (for example) to Edmund Storms's article in the May/June 1994 *Technology Review* (on which I commented in my January 1995 editorial), the U.S. Patent Office has categorically stopped issuing patents in that area, the Department of Energy has stopped funding research in it, and "conventional" journals refuse even to consider papers in it. Work in it continues, at what seems to be a large number of quite respectable labs, but has to be published in a very few special journals that are largely regarded as a "ghetto" and not taken seriously by those not in that ghetto. It may be that, if the effects they're studying are real, they'll eventually be heard even outside; but there is also a danger that the ghetto will be so effectively fenced off that a potentially revolutionary field will never have a chance to grow up.

That is why *Analog* occasionally publishes articles about such things. We are, at heart, a speculative magazine, and our readers include many scientists and engineers—some of them considerably above average in their openness to considering "heresies." Yes, it is important for us to tell you about some of the most important new work that is generally accepted. But we consider it just as important to tell you about some of the work that's having trouble getting heard in conventional professional journals, but looks as if it just might pan out and prove important. Much of it won't, of course; but some of it may. And someone who saw it in *Analog* just may play a key role in making that happen.

Advice for "Crackpots"

Since I do not want to get off on the wrong foot by antagonizing the very "crackpots" to whom these remarks are respectfully dedicated, I must begin by defining my terms.

A "crackpot"—in quotes—is an individual who is considered, as the dictionary says, "a harmless lunatic." A person, in other words, who expounds ideas or claims abilities that the generally acknowledged authorities agree are nonsense.

A *crackpot*—not in quotes—is an individual whose ideas or claims really *are* nonsense, in the sense that they contradict not the accepted wisdom, but the actual behavior of the universe.

Please note carefully that the two terms, defined thus, are quite distinct. A person may be either, or both, or neither.

The terms would be equivalent if the opinions of the generally acknowledged authorities were always exact descriptions of the real universe—but, of course, they aren't. They are approximations—sometimes very close, sometimes way off the beam.

And they change.

It's easy to come up with a list of people who were, at least to some extent, "crackpots" in their own times. Nicolas Copernicus and Galileo Galilei: the Earth goes around the Sun. Charles Darwin: species, including man, evolve from (and to) other species. Louis Leakey: man originated in Africa, not Asia. Albert Einstein: the speed of light is constant, but space and time aren't. Alfred Wegener: continents drift across the Earth's surface.

Not all of these fit the definition equally well. Some were not considered harmless; some had little trouble getting their work published, though it was controversial afterward. The actual difficulties they faced were seldom very close to the ones associated with them in the folklore, but each of them proposed ideas that were, at the time, at least mildly disconcerting to at least some people.

Not one of them is still considered (by most) a crackpot. Their ideas, once more or less widely scoffed at, have become part of the accepted wisdom—which has the interesting consequence that anybody who significantly *disagrees* with those ideas is now a "crackpot."

As such, somebody who challenges the theory of relativity or continental drift will, quite likely, have a harder than average time getting his ideas before the public (except, in very recent years, on obscure websites not likely to attract much serious attention). It may seem to him that there is a genuine conspiracy, with the world united to suppress his ideas, and *nobody* willing to give them a fair shake.

In general, this is an exaggerated view. Articles questioning generally accepted theories do get published, even in the more strait-laced professional journals. (An article questioning some detail of relativity or continental drift is, of course, more likely to be published than one questioning the approximate sphericity of our planet. This may have something to do with the age of the theories—but it may also have something to do with the amount and quality of supporting evidence that has been collected.)

Still, it must be admitted that a great many "crackpot" theories do *not* get published, and it is no doubt true that in many cases they are rejected at least partly because of prejudice. The disturbing possibility does exist that among those rejected theories are some that are genuinely better than their currently respectable counterparts, and that the failure to publish and pursue them blocks real and desirable progress—in some cases, perhaps, permanently. After all, the names above are all associated with what most of us now consider monumental advances, yet those advances were not greeted with chorused enthusiasm as soon as they appeared. . . .

For such reasons—and because really fundamental breakthroughs, by definition, *cannot* be deduced logically from previously existing theories—some of us think it important that some outlets exist where highly unorthodox theories can be aired. *Analog* has long been such an outlet, having carried extensive discussions of such subjects as dianetics, psionics, Dean drives, and neurophones. Some of what we've published has not held up. That's fine; we've learned something. In many other cases, important questions remain open. That's fine, too—provided somebody keeps trying to answer them.

But neither *Analog* nor any finite number of publications can possibly accommodate *all* the "crackpot" ideas that may contain something worth investigating. There simply isn't room. Therefore most of the unorthodox proposals *have* to be rejected.

And the blunt fact is that many "crackpots" really are *crackpots*. I don't mean to call anybody names, but a great many unorthodox papers that cross my desk really do contain blatant nonsense—ideas that are seriously at odds not just with fashionable theory, but with well-established *observation* of what really happens. Above I offered a brief list of "crackpots" who were eventually welcomed into the fold; I could make a much longer list of others who weren't, but it would be harder because most of them have been forgotten—in most cases, I suspect, deservedly so.

The real crackpots are the ones who give "crackpots" a bad name. And they make it harder for those with good, important ideas to be heard, in a still more basic way. If those ideas are to receive any support, encouragement, or constructive criticism, they must be shown to other people—research scientists, teachers, editors, etc.

And those people have only finite time to devote to such things.

Thus any serious "crackpot" must understand, at a very deep level, these two basic facts: (1) Anyone you send your work to will have seen a lot more genuine crackpottery than good original ideas among the unsolicited unorthodoxy that has crossed his desk, and this experience will have conditioned him to approach such things warily. (2) On the average, he will not be able to spend much time on such offerings. He may gladly do so if he sees immediate evidence that it may be worthwhile, but not otherwise.

I've had plenty of such things sent my way, both as a physics professor and as *Analog* editor. Among them were a few intriguing ideas—and a lot of gibberish: stuff that simply ignored a great body of experimental observation and added none of its own. Many were very long and not clearly written; in a few extreme cases I read several pages carefully and still had not the remotest idea of what the subject was. Some were single-spaced or even handwritten, which admittedly has nothing to do with the validity of the content, but certainly does have something to do with its accessibility. An appreciable number were accompanied by boasting or semi-threatening letters: "Put your money where your mouth is, sir. If you have any guts at all, you'll print this—but I realize it *will* take courage." (Well, that may be—though in some cases "gall" might be a better word. It would also take *space*, and I only have room for about a dozen fact articles a year. If you want yours to be one of them, *you* have to convince *me* that it deserves to be.)

Sometimes I'm not the first to receive these offerings. They arrive with letters plaintively recounting how others have refused even to look at them—plus, ironically, sheafs of correspondence showing clearly that others have, in fact, given them a great deal of time (which they must afterward have regretted wasting).

I look at *all* of these things. I try to see whether each really does have something new and important to say. I am *most* interested in new ideas, and there are few things more exciting than finding one. But experience has taught me that the probability of this time being productively spent is disappointingly low. Too often, when I have spent a lot of time on something like this, deciphering poor English and worse mathematics, I have found that the author simply did not understand the field. People try desperately to disprove relativity with no understanding of what relativity really says. Elementary mathematical errors totally invalidate conclusions. This sort of thing has happened so often that I now have a pretty good feel for when I'm dealing with this kind of material. This doesn't mean that I'll ever dismiss a piece out of hand, but it does mean that the time I'll spend on it is limited. I have other work to do, and, like it or not, I must assign priorities.

The situation of others, in industry or academia, is not so different. So what can you do if you have a good but unorthodox idea? I can guarantee nothing except that you *will* meet prejudice in some circles, no matter what you do. You will meet other resistance which is *not* prejudice, but genuine professional standards of a kind absolutely essential to science. But you will also find listeners, if you look hard enough and well enough—and your search may well have a better chance if you understand the following suggestions and keep them constantly in mind.

1. **Do your homework.** Learn what's been done (and this requires that you know how to do a literature search). If you're questioning an old theory, be sure you understand—thoroughly—what that theory says. Find out also what observational evidence has been accumulated for and against it. A theory such as relativity does not gain wide acceptance because of somebody's whim, but because it fits a large body of experimental data better than its predecessors. Basic breakthroughs will not be made by people who are sure the accepted knowledge is the last word, but they will not be made by people who are ignorant of what's been done, either. A healthy disrespect for authority is a fine and useful thing. A casual disregard of observed reality is not.
2. **Don't try to intimidate your reader.** I'm not sure what people hope to accomplish with those letters. Do they want their work judged solely on its merits, or not? The letters unwittingly suggest otherwise.
3. **Try to see your work from your reader's viewpoint—and understand that the time he can spend on it is limited.** When I, or someone else, receive an exposition of your experiments or ideas, we naturally hope it will contain something valid and earth-shaking—but, statistically, we know that's unlikely. I am not likely to

read a hundred single-spaced pages of abstruse prose, searching intently for a gem, if I don't find some indication quite early that this has a reasonable chance of being gem territory. This means two things for you as you write:

(a) Somewhere in your very first page you need to spell out—*in very clear, easy-to-read English*—exactly what is the essence of what you've done—and why it is important. Professional journals usually require an abstract in addition to the main text for this purpose. That never hurts, but in any case you should bear in mind that if I read two pages of your paper and still can't figure out what its point is, I'm not going to be very anxious to spend much time reading the rest. On the other hand, if I can see immediately that it does claim to make an important point, I'm going to be *very* interested in reading on to see whether you have solid support for that point. And if it continues to seem that you do, I will spend as much time as necessary to make sure of that.

(b) You need to make the *entire* text as concise, clear, and easy to follow as possible, aided where appropriate by clear and well labeled diagrams. Then if you've succeeded in point (a), so I want to read on, I'll retain that interest to the end—and be able to get through it all in a reasonable time.

If these sound suspiciously similar to recommendations you've heard for *any* kind of writing—that's no coincidence.

4. **Don't expect people to publish or research your idea unless *you* convince *them* it's to *their* benefit.** I, for instance, am in the business of producing a magazine for sale and enjoyment, and my publisher would not take kindly to my devoting many pages to material that sent my readers elsewhere feeling that their intelligence had been insulted. If you have an idea for a space drive, it's *your* responsibility to build a working model for demonstration—unless you can give a theoretical justification

so convincing that some company will consider it a good investment to do so themselves. Remember, it's *their* money you're asking them to spend.

5. **Don't waste time on indignant follow-up letters if somebody who sees your work does not respond enthusiastically—or at all.** Okay: he didn't see the merit in your masterpiece. That may mean that there isn't any, or that he's prejudiced and closed-minded (in which case he isn't going to be much help to you anyway), or that you failed to present your case well enough. It's always worthwhile to see if you can make significant improvements in the light of the suggestions above. If you can, it may be worthwhile to offer somebody a second look. Otherwise, just keep searching. Another type of follow-up letter may be in order, incidentally, if you're lucky enough to find two people interested in your work. If one of them manages to establish something new about it, by way of either support or disproof, it's only courteous to let the other know.

It should be obvious that I am interested in seeing good, new, offbeat ideas. If I weren't, I wouldn't have bothered to write this. When this advice is followed, the interchange of ideas can be potentially important or, at the very least, a lot of fun.

For example: A gentleman in Greece long ago proved to most mathematicians' satisfaction that angle trisection—the purely geometrical division of an angle into three equal parts using only a compass and straightedge—can't be done. Much more recently, a gentleman in Michigan thought he had done it. He had a construction worked out, as well as a working model of a device for carrying out a streamlined version of it mechanically. He did not have a rigorous proof that the construction was really what he thought it was, but he did have a detailed description of the construction which he was sending around in hopes of getting help in proving that it was (or wasn't) the construction so long thought impossible. He met the usual quota of glib rejections, but I was interested. He had a clearly defined problem and a description of his work which I could follow. It took much longer than I usually spend on such things, but it seemed worth continuing as long as I could not find a step that was obviously invalid—which was all the way to the end. I could see a way to determine rigorously whether or not the construction was exact, but I could also see that carrying it out was going to involve complicated and time-consuming trigonometry.

Meanwhile, someone else—an academic mathematician, as I recall—also got interested and carried out a detailed analysis before I found time. He found that it *was* not a rigorous construction, but it was an excellent engineering approximation, never off by more than a fraction of a degree. I really appreciated the inventor's taking time to write and let me know of that result—including sending me a copy of the professor's analysis before I had spent a lot of time duplicating it. We have had a highly enjoyable correspondence since then, and I look forward to more of such in the future.

But, in all honesty, I *don't* look forward to impenetrable tomes heralded by annoyingly defensive letters and expounding ideas which are either unintelligible or clearly and demonstrably untenable. By all means send me your ideas; I want to see them—but, please, only *after* you've assimilated the advice above.

Primitive Machines

You know the stereotypes. Perhaps they were best exemplified by the "Robot Theater" segments on television's old *Laugh-In* show, in which actors and actresses moved stiffly and jerkily about the stage, reciting romantic dialogue in a flat monotone and not quite managing to connect for the final clinch. But you've seen plenty of others. You've lost count of the cartoons, movies, or television shows you've seen in which computers or robots had to be addressed via keyboard in an esoteric, highly specialized language, and fouled things up by taking everything very, very literally. In some cases they spoke, but usually in that flat monotone; and if they responded to human speech, it was still with that extremely simple-minded literalness.

Written science fiction, of course, has fairly often been a little more imaginative; but the images in the preceding paragraph pretty well sum up the popular picture of human interaction with "smart" machines. The gist of it is: they may be smart for machines, but they sure are inferior to us humans, yup, yup! That attitude has found its way into our very language in a variety of ways. For example, where I spoke of a "flat monotone," many people would automatically say a "machinelike monotone."

Certainly there is a germ of truth in these images. By now practically all of us have actually heard flat-sounding synthesized voices. We have used, or at least attempted to use, computers with which we could only communicate by esoteric commands punched on cards or typed on a keyboard and bearing no discernible relation to our own speech. We have been frustrated by computers which completely misconstrued our intents because of a "trivial" misplaced comma, or insisted on interpreting commands in a bizarre way that was not at all what we thought we were saying. Many of us have been heard to mutter, "Stupid machine!" or words to that effect. It is not without reason that some computers bear signs hand-lettered by their operators, warning, "I will do what you say!"

But there is a subtle, profound fallacy in the popular view. It puts the emphasis on the wrong word. When we find ourselves thwarted by a "stupid machine," many of us tend to blame the problem on the fact that our adversary is a *machine*. But that's not the problem, really. The problem is that it's *stupid*. The fallacy is the assumption that stupidity and machinehood automatically go together. The fact is that stupidity and *primitiveness*—a low level of development—go together.

We have so far dealt only with relatively primitive machines—because we haven't been doing this for very long. We tend to forget that they're primitive, because they're the most advanced machines humans have ever seen. But they are, and we wrongly blame their shortcomings on the fact that they are machines rather than the fact that they are primitive. It would be a good habit to kick, since machines are now evolving quite rapidly. People who insist on feeling smugly and categorically superior to them are in for some *big* surprises!

Consider that matter of voices and the stiff, uncoordinated motions of the "Robot Theater" characters. They're funny, and they do bear a recognizable resemblance to at least some of the robots we've seen so far. The first synthesized voices *were* flat, monotonous, and devoid of inflection; and the first mechanical manipulators were stiff, jerky, and awkward compared to a human surgeon or ballerina. It would be astonishing if they weren't. The actions being mimicked are vastly more complicated than the humans who do them "automatically" realize, and the first attempts to duplicate them cannot be expected to get more than the basics. *But that's not the end of the line!*

If somebody actually wanted to build robots that resembled humans in appearance and behavior, enough to spend the time and money it would take, some of the early models might indeed resemble the caricatures on "Robot Theater." But if interest and funding held out—as they very well might, given the encouragement of even that much success—later models could be far more sophisticated. The television caricatures, whether their creators realize it or not, are not mocking the artificiality of robots *per se*, but rather the simplicity of *primitive* robots. And that can change.

In the case of voices, at least, it has already changed a great deal. In his February 1969 *Analog* editorial, John W. Campbell, fired by enthusiasm from a recent visit to the Bell Telephone Laboratories (a visit to Bell Labs will do that to you!), commented as follows on attempts to enable a computer to speak with a human voice:

"I heard a computer counting 'One . . . two . . .' not by playing back bits of recorded tape, but by generating the sound-forms of human speech. They were not monotonous, utterly impersonal tones, either; like a human being, the words could have a declarative tone, an inquisitive, or neutral tone. It could, in effect, say, 'Two?' or 'Two!' or simply 'Two.'

"True, as yet it can't count even to ten; as yet, while it can sound like a man's voice, it can't shift to the female mode, or to a small-child tone. But it's pretty clear that science-fictioneers have been wrong in suggesting that a computer's voice will be impersonal. By no means! They've already got a means of making it quite 'personal-human' in timbre."

That was big news, twenty years ago. These days, you probably hear synthesized voices so often you no longer think of them as remarkable, and in many cases you don't even realize that what you're hearing *is* synthesized. I have heard one singing the Queen of the Night's aria from Mozart's *Die Zauberflöte** so realistically that a listener who missed the introduction had not the slightest suspicion that she was hearing anything but a very good, completely human coloratura soprano.

I have also more recently visited Bell Labs myself, and met a

---

*A singular appropriate choice of repertoire for such a demonstration, if you think about it. A "magic flute," indeed!

robot named Sam who not only speaks in a realistic voice with normal inflections, but carries on real-time conversations in complete sentences, acts on instructions given in ordinary speech, and asks for clarification if he doesn't understand them. The very definition of "artificial intelligence" has evolved. The Bell researchers in robotics (or, as they prefer, "interactive systems") no longer consider it sufficient that a machine be able to manipulate lots of data. Now an "artificial intelligence" must be able to make sound decisions and take appropriate actions on the basis of incomplete and flawed data about a world which is constantly changing.

As we do. Sam still doesn't do it anywhere near as well as we do, or in as wide a range of contexts. But he's young. Give him time to grow up. . . .

Many people, of course, don't *want* Sam to grow up. They're scared to death of the idea that machines could *ever* be anything but primitive. Please note: I'm not talking about the people who actually know something about computers and related fields and have serious, thoughtful disagreements about the extent to which machines "can think," and exactly what such terminology means. I'm talking about the multitudes who, with little or no such knowledge, have a powerful, purely emotional reaction that says, "Obviously machines can't think, and never will!" I have asked a number of such people why they believed that, and found little more than a powerful need to feel clearly and forever superior to anything artificial. I think what I was seeing was fear. People who have never had to deal with a man-made being that could interact with them on more or less equal terms simply cannot face the potential dangers implicit in the possibility—so they simply deny that it exists.

There is room for legitimate debate about the extent to which "artificial intelligence" is a meaningful term, and the actual limitations of machine capabilities. You can read such debates in the works of such researchers as Marvin Minsky, Hans Moravec, and Roger Penrose. But regardless of whether we can achieve something you're willing to call true artificial intelligence, the evidence I've seen strongly suggests that we can and will have—and sooner than you think! things which are awfully hard to tell from it. They may or may not come in packages resembling us; but a great many lines of research are converging on machines which can manipulate, converse, make decisions, and learn from experience at a level of sophistication far beyond what many people are psychologically prepared for.

Does this mean that they won't be allowed to happen, or that humans will be unable to adapt to them? Neither of those seems very likely. The evolution of knowledge has such momentum that I don't realistically see fear dealing it more than a temporary setback. And I think the fear itself will evolve into a more useful sort of caution, in the usual way.

I remember when computers first began to be used as "teaching machines," and critics frantically warned of the "dehumanizing" effects of letting machines supplant human teachers. Many older adults are still afraid of computers, shying away from using them themselves, and more than a little distrustful of letting them get their electronic clutches on children's minds. But a generation of children is quietly growing up considering computers as normal and comfortable a part of the landscape as rocking chairs, getting complementary parts of their education from human teachers and from computers. From among those children, I think, will come the people best able to deal realistically with both the helpful and the dangerous potentials of highly sophisticated machinery.

Which doesn't just mean computers, of course. There has always, I suspect, been a tendency for people to judge the potentials of any technology in terms of its present capabilities, forgetting that today's "latest thing" is always tomorrow's "primitive and quaint," and fearing to look beyond today's horizon. It would have been a mistake to assume that people could never fly to the Moon because the Wright brothers' first flight could only stay aloft for twelve seconds. It would have been a mistake to assume that attempts to record music were a waste of time because the first efforts sounded terribly tinny, or that I could never have a multi-megabyte computer on my typing table because vacuum tubes were too bulky and hot.

Yet we still make comparable mistakes all the time, attributing shortcomings of primitive machines to the fact that they are machines rather than the fact that they are primitive. We persist in believing that robots will *always* be clunky and simple-minded because that's the only kind we can build *now*, and that electronic music will never sound like "real" music. Some even think they would like to get rid of as much technology of *any* kind as possible because it's energy-expensive and polluting. It doesn't occur to them that energy waste and pollution might be characteristics of primitive industry rather than industry *per se*. If their memories were just a little longer, they might remember what things were like in this country just a few decades ago, when soft coal was widely burned. (For a contemporary Bad Example, see eastern Europe.) The fact is that we have only recently begun to put much effort into making industry truly efficient and clean—and it would be a mistake of the classic kind to underestimate what can be done if that effort is supported and sustained.

Realistically, I suspect, even our descendants will continue to show some of the same sort of shortsightedness. But with change coming as fast as it now is, I hold out just a glimmer of hope that we finally may produce a generation, in the not too distant future, which actually *realizes* that change is a normal part of life—and that it doesn't always stop at the present.

Literature, Art, and Technology

Nouveaux Clichés

Not long ago I had a conversation with a promising new writer who mentioned that she, like many such, often meets with a group of other writers to "workshop" stories. At one point she mentioned that her writers' group sometimes criticized her work for a rather odd reason. "Why are you writing about these happily married people?" they asked. "That's such a cliché!"

"But it isn't," she objected. "Nobody else is doing it!"

Which is not *quite* literally true, but it *is* painfully close.

In rebelling against the clichés of *Leave It to Beaver* and *Father Knows Best*, it seems, many of today's writers have created and adopted a whole new set of their own. Unfortunately, not many of them seemed to have recognized yet that that's what they've done.

Marriage is just one example. Pick up a randomly selected story published almost anywhere in the last few years, and if a character is married, you don't have to read much more to be pretty sure that the marriage is falling apart. You *could* turn out to be wrong; but if you're a betting person, you won't find many bets with more favorable odds.

It's also an excellent bet that most of the characters will have more or less major psychological problems, and not much skill in dealing with them. The total cast of characters will very probably represent a mixture of races and genders in just those proportions deemed Correct by Popular Wisdom circa 1993. Their concerns and attitudes will closely mirror the set of contemporary fashions known by that odious label "Politically Correct."

All of these things have become so commonplace, so *predictable*, as to easily qualify for full-fledged cliché status. As Robert Coulson remarked in a recent review, ". . . there was a revolution in writing styles and editors today want stories based in the 'real world' and characters who are 'real people' (spiteful, psychologically immature, sexually confused, self-absorbed, etc.)."

But how did it get decided that such people are more "real" than those who *can* manage their own lives, enjoy them, and be a source of joy to those who know them?

I flatly reject such a claim as faulty observation, because I personally know quite a few people of the sort I've just described. Yes, of course I know quite a few of the kind Coulson describes, too. But a literature that populates itself exclusively with those, and ignores the others, is every bit as clichéd as *Leave It to Beaver* or *Space Cadet.*

"But," you may object, "even if it's true that there are some happy, competent people in the real world, they don't meet the needs of Literature. They're not *interesting*!"

Au contraire, say I. If happy, competent people are as scarce in reality as the current crop of Literary folk would have us believe, I should think they would be *very* interesting—as oddities, if not as role models or founts of inspiration. And if they're that scarce in reality, there must be lots of readers who could *use* them as role models or founts of inspiration.

This assumes, of course, that writers are capable of creating them. (Could *that* be the problem? I dare you. . . .) I'm not suggesting a wholesale return to the simple-minded Good Guys and Bad Guys of yore; there were good reasons for writers, editors, critics, and audiences deciding they'd outgrown those and it was time to move on to something else. I'm suggesting that now we've outgrown the present state of affairs and it's time to move on still further.

In deciding it was time to give more attention to Characterization, I think, many writers and critics moved from one clichéd extreme to another. Once upon a time Good Guys had no faults, which is a very rare condition among human beings. Now hardly anybody has any virtues—which is also rarer than many would have us believe. It has become commonplace—even *de rigueur*, in some circles—to confuse weak characters with strong characterization.

But strong characterization is not as simple as peopling stories with individuals who have made messes of their lives and can't figure out how to unmess them. It means creating a *variety* of characters who look and sound and feel so believable that a reader can imagine them, and be interested in them, as real people. (Note, by the way, that in general readers do *not* want realism, but *verisimilitude*, which is an entirely different thing. If you're not clear on the difference, look it up—and then think about it.) Naturally those characters will have problems; problems, and people's efforts to solve them, are the very essence of *story*.

Some of those problems will even be psychological. *Analog*'s readers, and I, are certainly interested in sympathetic portrayals of such people, provided they make the best effort they can to *do* something about their problems. But I would also like to meet an occasional someone in fiction who is so vividly portrayed that he or she seems real, yet whom I might like to know, and perhaps even admire and emulate, in real life. I am *not* interested in a steady diet of neurotic incompetents who make no effort to become something better. Look at it this way: if you're a writer, your characters are people I choose to spend my time with when I read your story. I'd like at least some of them to be people whose company I might seek out in real life.

There's at least one other thing that needs to be said in response to the worn-out convention that stories can't be interesting unless all their important characters are psychological disaster areas. Such a belief implies that the universe contains nothing interesting enough to write or read about except corrupt and/or unbalanced minds and souls. Some people *do* believe this; William Faulkner is often quoted as saying, "The only thing worth writing about is the human heart in conflict with itself." But Faulkner was wrong. That's *a* thing worth writing about, and obviously one of special importance and interest to human beings. But to say that it's the *only* thing worth writing about betrays a dangerously narrow, anthropocentric, perhaps even arrogant attitude.

One of the fundamental precepts of science fiction—*real* science fiction, not thinly disguised mainstream fiction or fantasy that has misappropriated that label on some flimsy pretext—is the recognition that the universe contains an enormous variety of interesting and important things. "The human heart in conflict with itself" is important to us because we *are* human beings; in the cosmic scheme of things, it's small potatoes. Since we normally see things from a human rather than cosmic perspective, human problems and conflicts are, should be, and will continue to be an important part of human fiction, including science fiction. The most memorable fiction usually has a strong impact on the emotions as well as the intellect.

But people who have anything resembling a realistic grasp of the scope of the universe and our place in it will find a great many other things worth exploring. Sometimes those will even be the most interesting and important things in the story, and won't need to be shored up by bogging their protagonists down with miserable childhoods and failing marriages. Not only is it not necessary for every story to deal with the hidden recesses of badly warped or damaged psyches, it's not necessary for *every* story to put its primary emphasis on people at all.

If you think about it, you'll realize that that, too, is in keeping with real life. You neither need, want, nor have any occasion to learn about the innermost thoughts and hangups of most of the people you come in contact with. And if you're accosted by an armed robber in the midst of a skydive, you're not going to have *time* to psychoanalyze each other—but the purely external conflict can generate one heck of a story!

So how about we try to get a little more perspective back into our storytelling and our values for judging it? I pose three simple challenges:

1. **Don't confuse weak characters with strong characterization.** Give me interesting people to read about, but don't kid yourself that it's either necessary or sufficient to make them all seriously neurotic or unprincipled in order to do that. I've seldom liked

a story unless I liked at least one person who lived in it. If you're telling me a story, I'd like to meet somebody in it who I remember as real and worth knowing. Intelligence and integrity are *not* dirty words!

2. **Remember that human beings are not the center of the universe, or even the only interesting thing in it.** Yes, I want to meet interesting, memorable people in your stories—but please don't burden either them or me with tedious details of their personal hang-ups if they have no particular relevance to the events they're caught up in and those events are themselves more interesting. And I hope that sometimes what's happening around the characters *will* be at least as interesting as the characters themselves.
3. **Dare to be fun.** Now that science fiction has become Respectable, many writers, egged on by Serious Critics, seem to be so busy trying to be Profound and Meaningful that they forget that the *first* purpose of all this is to *entertain.* This is not to say that everything in fiction should be sweetness and light, or that every story must have a happy ending. One of my all-time favorites is Daniel Keyes' "Flowers for Algernon," which would have been utterly destroyed by a happy ending. But a worldview, in fiction or in life, that is all bleakness and *Angst* is just as distorted as one that includes none of these things.

A reader once wrote that she was cancelling her subscription to *Analog* because it failed to reflect her view that the world is a terrible place. I had to reply that it couldn't, because the world I've seen is *not* a terrible place. Or perhaps I should say that it is, but that's only one of its many, many aspects. The universe is a terrible, wonderful, frightening, soothing, sad, funny, ridiculous, sublime, depressing, exhilarating place, and it's probably going to keep getting more so. I want my fiction, whether I'm reading it, writing it, or publishing it, to reflect all those facets, not just whichever of them some individual happens to fixate on. I hope you as readers will demand nothing less of those you read.

The Dark Side of Clarke's Law

Probably everyone who's been reading science fiction for even a moderate time is familiar with the observation often called Clarke's Law, after author Arthur C. Clarke: "Any sufficiently advanced technology is indistinguishable from magic."

What Sir Arthur originally had in mind, I suspect, was that an alien culture—or a future human culture—that had had much more time to develop might use technology whose operating principles would be so unobvious to us that they might as well be supernatural. By the same token, our own technology might look like magic to members of a culturc that had remained isolated and at a Stone Age level of development somewhere on Earth. Anthropologists have had a few occasions, such as first contact with the Tasaday, to verify that this actually happens.

Until recently, I tended to think of Clarke's Law as simply an academic observation about what was likely to happen if we ever encountered a civilization at a very different level of technological development. It didn't seem to have any particular practical significance, or beneficial or detrimental consequences, unless and until that happened. Recently, however, it has occurred to me that Clarke's Law *may* have practical consequences—not necessarily good—and much closer to home.

This realization hit me in January 1996, when I was being interviewed by Canadian journalist Joe Woodard for an article he was doing for the magazine *Alberta Report*. His article ran in the January 22 issue, under the title "Space: The Forgotten Frontier." His starting point was the observation that planetariums have been losing popularity, but that's just the tip of the iceberg—just one symptom of a much more widespread fading of interest in, and support for, space exploration in particular and science in general.

Mr. Woodard asked my opinion about whether there was a corresponding drift of popular interest away from science fiction—*real* science fiction, in which the word "science" actually means something—toward fantasy. That one was easy. It's been clear for some time that fantasy, labelled as such, has acquired a very large share of the "science fiction and fantasy" market. Even much of what's now labelled "science fiction" has little interest in scientific rigor or plausibility. The kind that does—the kind *Analog* has always tried to provide—has come to be regarded in some circles as a rather odd minority faction, fenced off in its own little ghetto and posted with the warning sign "hard science fiction."

Then Mr. Woodard asked me, "Why do you suppose this is?"

And it suddenly occurred to me that this phenomenon just might be a case of Clarke's Law in action, not in the way one culture sees another, but right here within our own culture.* To many people who use it, *our own technology has become indistinguishable from magic*! We all use many everyday things, such as personal computers, television, and microwave ovens, that would have seemed like pure, unmitigated magic to most of our ancestors. Well, they seem like magic to many of us (present company excepted), too—because many, perhaps most, people use these things with little or no understanding of how they work.

The practical consequence? If too many people view existing technology as "magic," done by a sorcerer class fundamentally different and distant from themselves, too many of them may be content to use what they have and not be motivated to learn how to further advance it.

There are at least two aspects to this blurring of popular perception of the distinction between science (and its applied offshoot, tech-

---

*Though C. P. Snow might say that those who do science and those who just use it *are* two different cultures!

nology) and magic. First, I think there's a growing tendency to feel that we already have "magic" to do so many things that there's little need to concern ourselves with discovering or inventing more. Second, there's a tendency to feel that science is so incomprehensible that "there's no point in *my* trying to learn about it because it's too hard." Certainly there's always been a fair amount of that feeling around, but it does seem to be particularly potent now. (If you need more convincing of just how abysmally ignorant yet sneeringly disdainful many American adults are of real science, you might be interested in an essay on "Science and America" by Nobel Laureate J. Michael Bishop, in *The Gettysburg Review*, Volume 8, Number 4, Autumn 1995.)

The blurring in the popular mind of the distinction between science and magic makes an obvious parallel to the blurring of the distinction between science fiction and fantasy. That parallel is not just an academic curiosity. There has long been a symbiosis, a mutual feedback, between science and science fiction. Science fiction has built on ideas from science—and in return has helped *create* new science by firing up the interest that led to scientific careers. Isaac Asimov, both a scientist and a science fiction writer, once estimated that half of all creative scientists were drawn into their fields at least partly by an early interest in science fiction.

Both science fiction and fantasy can be fun. What has made science fiction special is the idea that its subject matter is not only fantastic, but at least marginally *possible*. Few people read stories of dragons and elves and set themselves the career goal of *becoming* a dragon or an elf. But a great many people, in earlier decades of this century, read tales of space travel, computers, or medical advances, recognized them as real possibilities, and told themselves, "I'm going to help make it happen!"

And they did.

But to do so, they had to recognize a few other things, too. They had to recognize that turning such dreams into reality requires a solid understanding of definite physical laws. They had to recognize that gaining and applying that understanding takes *work*. And they had to recognize that, challenging though it is, such work is not beyond the capabilities of real people. Someone with enough smarts to see the possibility, and enough determination to learn what it takes, *can* help make it happen.

But how often will that occur if more and more people become less and less able to grasp the distinction between science and magic, or between science fiction and fantasy? If everybody sees the technology already achieved as "magic," provided by Somebody Else to be used without understanding, who's going to create the next generation of it? (Or, for that matter, keep the old operational?) If readers fail to see science fiction as differing from fantasy in that it shows real possibilities that they themselves might help realize, how many will be inspired to do so?

It *may* be a false alarm—fads come and go in intellectual attitudes as in everything else—or it may be a real problem. If it *is* a real problem, what can be done about it?

A variety of things, none of them necessarily easy—but, as with science itself, that is not a reason not to try. Science teachers need to use whatever tricks they can to make real science and its potentials fire young imaginations not only with interest, but with the confidence that they, too, can aspire to make real contributions. Not, I emphasize, just tricks for the sake of razzle-dazzle. There's nothing wrong with razzle-dazzle—John W. Campbell was right when he said fervently, "Teaching ought to have more *circus* in it!"—but educational circus has to do more than entertain. It's fine if it can do that, but its real purpose is as a means to an end: to make students learn, and want to learn more. And teachers must never fool themselves into thinking that just because they've gone through the motions prescribed in the curriculum, they've done their job. The acid test is: Did it work? If not, they must try something else, and keep trying until they find an approach that does work.

And we in *this* field must discourage the growing illusion that great things will be done by handwaving. Some of us must still take the trouble to write science fiction with equal emphasis and attention to both words of the name and to label our work carefully and honestly. How are people to see science fiction as something different from fantasy, something dealing with real possibilities just waiting for someone to do the work to make them real, if much bearing the label "science fiction" *isn't* that? How are people to understand that important complex of distinctions when something based on a vague faith in "The Force" is billed as the same kind of stuff as the visions of Clarke, Heinlein, or Asimov?

It has become fashionable in ever widening circles, *even among writers of science fiction and fantasy*, to pretend that there's no difference between them. It's true that some stories fall into a rather hazy "twilight zone" on the border, and a case can be made for calling them either science fiction or fantasy rather than clearly one or the other. But, as Kelvin Throop astutely observed, "Just because there's twilight doesn't mean you can't tell the difference between night and day." Encouraging people to pretend there's no difference does no one any favors, and just may have a real effect on how our culture develops—or fails to—in the future.

For starters, I'd like to see the phrase "hard science fiction" phased out—and replaced by a more careful distinction between the arts of science fiction and fantasy. "Hard science fiction" is too often too narrowly interpreted, and too often scares away readers who interpret the word "hard" in the sense of "difficult," even if writers intend it to mean "rigorous." All it really means, as I interpret it in my own fiction writing and in editing *Analog*, is this: Some scientific or technological speculation is so *integral* to the story that it can't be removed without making the whole story collapse; and whatever science the story uses is made as *plausible* as possible (whether by strict extrapolation of known science or by imagining new principles that might be discovered in the future).

Any science fiction in which the term "science" has any real meaning meets those criteria, and anything that doesn't is not *science* fiction, but something else. So we don't really *need* the term "hard science fiction"; it's just an unnecessarily severe synonym for *science* fiction.

But we do need a generally understood distinction between science fiction so defined, which by its nature offers real possibilities, worthy at least of real-world investigation, and fantasy, which may offer top-notch entertainment but makes no pretense of offering real possibilities. If we can just get that distinction across, our field may yet do its part to spread understanding of the fact that doing real science takes real work—but it's work that can be done, and it's the only way we have to make new kinds of "magic" real.

Sensory Deprivation

When I was in graduate school becoming a physicist, we had weekly colloquia by visiting speakers from other institutions. One of them was a Nobel Laureate; I don't remember *which* Nobel Laureate, or what his main topic was, but I do remember the story with which he opened his presentation.

It seems there was a professor doing research in neurophysiology who finally achieved a breakthrough late one night and was so eager to share his triumph that he collared the night janitor to show off his achievement. "Look," he crowed, "I've managed to capture all the knowledge and wisdom and experience of one human being in a chemical extract that can be injected into another person and give him the same benefits. People can be whatever they want to be, and by selling the potions that make it possible, I'll be a rich man. Look at these jars. . . ."

The janitor followed the professor's pointing finger to a shelf filled with neatly labelled jars: "Essence of Chemist's Brain—$2,000/oz. Essence of Mathematician's Brain—$2,000/oz. Essence of Philosopher's Brain—$2,000/oz. Essence of Physicist's Brain—$35,000/oz."

"That's really something," the janitor granted. "But tell me—why is the physicist's brain so much more expensive than the others?"

"My dear fellow," the professor asked patiently, "do you have any idea how many physicists it takes to get an ounce of brains?"

The audience—consisting almost entirely of physicists—laughed enthusiastically. At least most of them did; when I consider my observations of the population at large, I wouldn't be surprised to find out that there were a few there who found it offensive and embarrassing that a Nobel Laureate would waste even a half minute of his precious lecture time on a joke—and at the expense of his fellow physicists, at that!

I'm not sure that the number of people who frown on most attempts at humor is any larger than it's ever been—I can remember examples from quite early in my life—but I think it may be. I've been having the slowly growing impression that our society is going through a period of taking itself Ever So Seriously about practically everything.

Not everybody, of course. When *Analog* invited guest editor Kelvin Throop to produce a special spoof issue a few years ago, more people wrote to say they found it delightful than disgraceful. (Which was a pleasant surprise, actually, since people commonly only bother to write to editors when they're angry.) But there *were* a significant number of people who thought the whole concept was beneath such an august periodical's dignity, and even more who were offended by one or more specific items in it. This wasn't particularly surprising; I've noticed over the years that whenever we run *anything* funny, a few people can be counted on to register their disapproval. A fact article with humorous asides or footnotes will always produce a stiff reminder or two that, "Sophomoric humor has no place in a serious discussion of ______________." (*You fill in the blank.*)

Apparently all editors, at least of "serious" publications, find the same thing. Hardly a week goes by that my local newspaper doesn't print a letter demanding that Gary Larson's *The Far Side* or Bill Watterson's *Calvin and Hobbes* or Garry Trudeau's *Doonesbury* be dropped on the grounds that it's sick and disgusting. Nor are news and editorials exempt. A recent feature article started out: "The message is becoming so abundantly clear, even beer companies are getting into the act. Drunken driving is no longer funny—or excusable."

How many public figures would dare to attempt a humorous remark that even *mentioned* anyone's ethnic background, however innocuously, knowing that to some people no such remark could *ever* seem innocuous? And how often have you seen a humorous story, play, or movie win a major award?

Let me toss out a few quick responses to some of these observations before getting down to more methodical discussion:

1. **Some of us think that humor can have a place—an *important* place—in the discussion of just about *anything*.** If nothing else, it can help to attract and hold a reader's interest, and make serious discussion more palatable and therefore more likely to happen. Whether it's "sophomoric" is very much a matter of personal taste; so, somewhat regrettably, it won't work the same way for all readers. However . . .
2. **The proper solution for readers who don't like a particular comic strip is: *If you don't like it, don't read it.*** It's unrealistic to expect all readers of any publication to be equally pleased with everything it publishes. Just skip the parts you don't like, and kindly allow others with different tastes the same privilege.
3. **Drunken driving has *never* been excusable—but drunkenness can still be funny.** It is possible for the same intelligent, decent person to laugh at a Jackie Gleason sketch involving an intoxicated character and to campaign vigorously for stronger legislation to fight drunk driving. Humor can even be used as a weapon against that or other evils; but the refusal or inability to see humor in a situation can be self-destructive.

One of the items in Mr. Throop's aforementioned spoof issue of *Analog* was Laurence M. Janifer's short story "Love in Bloom." The story grew out of the author's whimsically changing the famous opening line of Robert A. Heinlein's "Blowups Happen" from "Put down that wrench!" to "Put down that wench!" and then concocting a plot to go with it. At least one irate reader apparently read the story as condoning rape, and wrote a letter to which the author composed a thoughtful reply that deserves to be pondered by a much wider audience. He has kindly given me permission to quote part of it here:

> "I don't consider rape or the threat of rape a casual and unimportant thing. I do consider it an improvement on murder, whether I get raped or you do, for the reason specified: post-rape, I'm alive. And I consider it, like death, insanity, and all other important things, material for humor. Nobody laughs at small items. People laugh mostly at big things, suitably presented. I like writing funny stuff and I like reading it, and I have no special worries about subject matter. At a few edges (the death of a child, for instance), the easiest form of humor is black humor, 'sick jokes' and the like. But *anything* is suitable material for funny treatment, I think. . . ."

And for a very good reason, too. For me, it's justification enough for some forms of humor that they can make necessary tasks more interesting and enjoyable. It's no coincidence that the very best academic course I ever took, in terms of what people *learned* from it, was also one in which humor was a prevalent part of virtually every class session. But what can possibly justify joking about serious matters like death and insanity and prejudice and war? Let Hawkeye Pierce (of *M.A.S.H.*) answer, as he answered when another character complained that he made a joke out of everything: "It's the only way I can keep from screaming."

Humor is an important mental safety mechanism. The sense of humor is a sense that human beings, individually or collectively, cannot afford to be deprived of. It's essential that we do try to solve problems, but the self-righteous sourpusses of the world must not be allowed to forbid us to laugh at them, too.

Humor is a unique phenomenon. As Arthur Koestler writes in the *Encyclopædia Britannica*, "Laughter is a reflex but unique in that it has no apparent biological purpose. . . . Its only function seems to be to provide relief from tension."

And a very important function that is.

Philosophers, psychologists, and physiologists have been trying for centuries to define and understand what humor *is*. It isn't easy, if only because it takes so many forms and is so different for different people. They seem to have done fairly well with identifying two common elements: (1) the perceiving of a single situation in two contexts or frames of reference, each internally consistent but not compatible with the other, and (2) some element of aggression. The latter may range from outright malice or contempt to a mere absence of sympathy for the victim of the joke.

This last might be compared to the "willing suspension of disbelief" that science fiction demands of its readers. For the duration of a joke, the listener must agree not to think too intently about how the person on the receiving end feels, or the story will become tragic or pathetic rather than funny. Thus kidding can't be too close to literal truth (satirists *must* exaggerate), and the humor of educated adults in civilized societies tends more toward the subtle than the malicious. Primitive hunters can laugh at the agonies of a mortally wounded prey animal because they don't have the concept of the animal being enough like themselves to inspire empathy. Civilized people do, so the aggressive element in their humor is more likely to be "so faint that only careful analysis will detect it"—or, as some might claim, far-fetched.

But what if it isn't? Might even coarser forms of humor have some value to humanity? Many people who have studied the problem have thought so. Aldous Huxley wrote, ". . . we tend to produce more adrenalin than is good for us, and we either suppress ourselves and turn destructive energies inwards or else we do not suppress ourselves and we start hitting people."

To which Arthur Koestler added, "A third alternative is to laugh at people. There are other outlets for tame aggression, such as competitive sports or literary criticism, but they are acquired skills, whereas laughter is a gift of nature, included in man's native equipment."

Humor provides an escape valve, a socially acceptable release for excess tension or hostile feelings. Given the choice, it is better to be laughed at by people who don't understand you than to be slaughtered by them. Is such a choice necessary? It would be nice to believe it isn't, but I'm not sure that's the way things really are.

It is true, as I've already mentioned, that humor has evolved. We are no longer amused by things that amused our distant forebears, because we have learned to empathize with more of our fellow beings. Is it not then all to the good that things change so that fewer things become funny? Isn't it a continuation of the evolution and refinement of humor that we've all become so empathetic that we now see foreigners and people of other races as fellow people instead of objects of humor?

It might, except that I don't believe that's what's actually happened, at least so far. Certainly it's true that *some* of us don't go in for ethnic humor because we've actually outgrown the need for it; but I fear that many avoid it because, and only because, they fear the consequences if they don't. And this may, in the long run, do more harm than good. It would be wonderful (or would it?) if we were in fact a society that had reached such a lofty peak of perfection that it no longer needed to laugh at anyone or anything. But it often seems to me that what we really are is a society that is trying desperately to delude itself that it has reached that peak before it actually has.

That in turn reminds me of the old cliché about closing the barn door after the horse is stolen. If a society decides to make it unacceptable to joke about certain matters, might it in effect be closing off the escape valve that used to release excess tensions, before the source of those tensions has been removed? If so, the ultimate effect could be very much like that of "fixing" a house with faulty wiring by eliminating its fuses.

Need I elaborate?

If individuals want to deprive themselves of the ability to laugh, at anything or everything, that is their business. I may feel sorry for them, but I do not consider it my duty or right to try to change them. The blacks and Poles and blind and physicists I've known who were able to joke about themselves as well as others have generally seemed much higher in self-respect and happiness than those who couldn't. I suspect that eventually, when we have actually outgrown our present terror of offending or being offended, we will find fewer people inclined either to practice the crueler forms of humor or to overreact to remarks in which no real malice is intended.

But in the meantime, I object strenuously to those who would try to deprive others of their freedom to joke. Humor is much too important to lose. As the interpreter Sakini says at the end of John Patrick's play *The Teahouse of the August Moon*: "Pain makes man think. Thought makes man wise. Wisdom makes life endurable."

And a sense of humor is a vital part of wisdom.

The Old-Timer Effect

There is a type of letter that I (and apparently every other editor) receive with some regularity, the gist of which is that science fiction (or some other field) has gone to the dogs and lost the magic that it had in the Good Old Days. Why, the writers of these letters ask, do the newer authors virtually never produce anything that can hold a candle to what the older ones did?

By a curious coincidence, a large majority of these letters come from people who have been reading the field in question for many decades. Not always; around the time I started thinking about writing this, I had recently collected several such letters from readers in their seventies or eighties—and one who had not yet hit thirty. Okay: there's no intrinsic reason why a person of *any* age should like everything new (*I* certainly don't!), or even prefer art or literature produced in his own period to that of one or more earlier times. But when large numbers of people, mostly of relatively advanced age, express essentially the same complaint—that Things Ain't What They Used to Be—in regard to a variety of fields, and when this has happened in every period of recorded history . . .

It gets awfully tempting to suspect the influence of some systematic phenomenon, something that tends to happen to human beings who live a long time and experience a lot—something that happens *as a result* of living a long time and experiencing a lot. I propose the name "The Old-timer Effect," and define it as the tendency for people to find it harder and harder to appreciate new things as they themselves grow older. The evidence I've seen suggests that it is real, common, and has a significance that goes far beyond science fiction, painting, music, or any other specific field of endeavor. For example, at a time when some scientists are talking seriously about the possibility of greatly extending human life spans, it might be prudent to ask: How many people would really *want* immortality, or near-immortality?

"The Old-timer Effect," I think, involves at least three closely related phenomena. A given individual may experience any one of them, or any two or all three in combination.

1. **People tend to specifically remember only the best of the old stuff, and to mistakenly think of it as typical of the entire period.** This is almost tautological; "good," in regard to art or literature, might well be *defined* as "tending to stick in the memory." It's hardly surprising that the "not-so-good" doesn't. A selective memory has definite advantages for a being who collects decades of experience and has no need for ready access to most of it. But that doesn't change the fact that the not-so-good (or not-so-memorable) was there—frequently in larger quantities than what *did* stick in memory. In the Good Old Days, I'm sometimes told, science fiction was full of classics like ________, where the writer fills in the blank with a number of personal favorites. Furthermore, he or she sometimes adds, there weren't all those typos that these uppity young copyeditors and typesetters and proofreaders let slip by.

   It never seems to occur to most of these writers that the examples of How Good Things Used To Be are the *most striking* examples they've gleaned from among many others that they don't remember at all. I can't go along with the illusion, because I have a complete set of back issues in my office, and have spent quite a bit of time exploring them *recently* in the course of compiling anthologies. I've *seen*, recently, the large amount of relatively forgettable material in which the gems are embedded (and I've also seen a good many typos, even Back Then). I've also seen it happen in other fields, such as the list of then-recent compositions being promoted by a major publisher of orchestral music on the back of a part printed some thirty years ago. I'd be very surprised and impressed if you recognized more than four of the 104 pieces or two of the thirty composers listed.

Occasionally the writer of a Good Old Days letter is perceptive enough to recognize that his memory of an old story may have been enhanced by the selectivity of memory or the fact that his tastes just weren't very discriminating when he first read it. Such a writer may strengthen his argument by adding that he has reread the old favorite and found that it was not only memorable, but held up well on rereading. But how much, and exactly what, does rereadability prove? It may reflect little more than the comfortable quality of familiar things with pleasant associations.

2. **The more things you've experienced, the harder it is to find new ones which are different enough from any of them to strike you as fresh and new.** Remember when knock-knock jokes were funny? To a child who's never heard one before, they really are. After the pattern has become familiar, and after you've learned to appreciate some others, it gets awfully hard to find a knock-knock joke that does much for you, because the elements of surprise and newness are gone. The same thing can happen with any kind of experience. You may continue to find jokes or symphonies or sports or people with characteristics that are new to you, but the more jokes or symphonies or sports or people you've known, the more new ones are likely to remind you of old ones. If your enjoyment of life depends on how different your new experiences are from your old ones, you are in ever-growing danger of finding more and more things boring.
3. **On the other hand, *many people, as they age, don't really want much novelty.*** The older they get, the more they fear the new and unfamiliar, and crave the old and comfortable. Instead of something fresh and new (even if they say they want that), they may really want more of what they liked best from their

> own past—which would not satisfy many younger people who really are looking for something new and different. Thus it is that virtually every new form of art, literature, or music in history has been resisted by older generations (and frequently touted by younger ones to an extent completely out of proportion to posterity's eventual judgment of it).

Again, I emphasize that I'm not denying that a decline in the quality of anything, measured by whatever criteria you choose to use, can occur. Certainly it can, and both producers and consumers need to be on guard against it. Nor am I saying that anybody is under any obligation to welcome all new trends, or to favor the cultural works of any period over those of any other. After all, if the function of such a work is to do something positive for the viewer or reader or listener, whether it does so is the only significant test of its value. The result of that test can and will be radically different for different viewers or readers or listeners, without meaning that anybody is "wrong."

What I *am* saying is that those who would analyze progress or decay in such subjective matters cannot leave themselves out of the analysis. They need to be aware that their memories tend to retain only what is "best" by their particular standards, and forget that the rest existed. That tends to make the present look "bad," whether it is or not, simply because they have not yet had time to forget the less memorable majority of recent work. They need to be aware that the more things they've seen, the more things will look familiar—which tends to make even new ideas stand out less than they used to because there are more "old" ones around them. And they need to be aware that they themselves may become less receptive to new ideas, and come increasingly to yearn for the comfort of familiar ideas, patterns, and styles.

Since it is possible to do all these things *without* being aware of them, the same person may experience both of the last two manifestations of the Old-timer Effect, even though they seem to contradict each other. That is, he may *think* he wants something fresh and new and be bored by his failure to find enough of it; yet when he does find it, he may find it disquieting and distasteful. That may make him wish for more like the Good Old Stuff; yet when he finds new work like that, he may complain that it is merely an unoriginal and unsatisfying imitation of things he has already read and remembers fondly.

None of this is intended as a personal attack on anybody who has experienced any of these tendencies. They are not "bad" or malicious; they are simply something that happens to people—and a rather sad thing at that, as anything that diminishes the enjoyment of life, whether internal or external, is sad. And probably none of us is completely immune.

Naturally there are exceptions. Some people are better than others at retaining a fresh outlook and finding new sources of satisfaction quite late in life. Some manage to do it throughout quite long lives. The composer Giuseppe Verdi did some of his most inventive and highly regarded work between the ages of seventy and eighty-five. "Grandma Gatewood," a legendary figure among Appalachian Trail hikers, walked the whole two-thousand-mile trail (again!) in her eighties. I have known several people personally who managed to keep exploring the universe, and reveling in what they found in it, well into their seventies, eighties, or even nineties. I hope to be one of them, eventually.

But I have also seen the Old-timer Effect in so many people that I must consider it almost an occupational hazard of being human. Sometimes it begins to take hold quite early. I have heard parents in their twenties say things like, "Having a child restored my joy in life," and I wondered privately why they had even begun to lose it that early. And while the prospect of very long life still sounds highly desirable to me, I must wonder how many people will really be able to enjoy how much of it.

Upstart Instruments

There's a common tendency to regard art and technology as quite separate and distinct. They're seen as the two contrasting halves of a sharp dichotomy, as in the cliché, "An art, not a science." Sometimes they're even seen as antagonistic, "artsy" types looking with disdain on anything as "cut and dried" or "nuts and bolts" as science or technology.

In reality, of course, it's not that simple. Every art depends on one or more technologies as its tools, and artists tend to embrace eagerly new techniques that help them get the effects they want more easily or effectively. Painters who once would have worked with oils and camel hair now are at least as likely to use acrylics and airbrushes. Writers who would have used quill pens when this country was new graduated a century ago to typewriters and a decade ago to word processors. Trumpeters are constantly trying new mouthpieces and leadpipes to improve their instruments' performance; clarinetists experiment with reeds and ligatures. Composers, like authors, have software to facilitate their work.

Yet there is resistance, too. Orchestras have long tuned to an A provided by an oboist, but in recent years pocket-sized electronic tuners have become commonplace and some conductors prefer them as their pitch standards. Oboists sometimes feel insulted by such a preference, feeling sure that they can provide at least as reliable a pitch as "that box." But the fact is that they can't. It's not their fault, and they shouldn't take it personally. The physics of the two systems simply dictate that there's no way a vibrating air column in a wooden tube can be as stable and reproducible in frequency as a quartz-controlled electronic oscillator. So oboists are probably going to have to resign themselves to being replaced more and more often as the pitch standard for orchestras.

A hot topic among professional musicians lately has been, "What can we do about the electronic upstarts that are horning in on our turf?" Union newsletters are full of articles and letters about the perceived threat to musicians' livelihoods posed by electronic instruments such as synthesizers and the growing prevalence of recorded, rather than live, music.

Those are really at least two separate issues. Keyboards and synthesizers are, at least in first approximation, simply new kinds of musical instruments. Viewed that way, it seems oddly ironic that some of the people who feel they must be protected from "unfair competition" from these newfangled contraptions play things like the saxophone (which was a newfangled contraption in 1846), valved brasses such as trumpets and French horns (circa 1815), and the piano (which was invented in 1709 and began displacing the harpsichord around the middle of the eighteenth century).

Some musicians realize this, of course. Along with the doom-crying letters and articles are others pointing out that this is by no means the first time new instruments with new abilities have come along, and that it might be more prudent to learn to take advantage of them than to bewail their existence.

Synthesizers, admittedly, are not *just* a new kind of instrument. What makes them threatening to traditional musicians, and not without cause, is the fact that a single synthesizer can, for at least some purposes, replace not just one or two instruments, but an entire orchestra. (Or chorus! Tchaikovsky's ballet *The Nutcracker* includes some passages for offstage women's chorus; I've played performances in which the "chorus" was a Yamaha keyboard and nobody I asked in the audience had the slightest suspicion that it was anything but a women's chorus.) Already a very large percentage of music for radio and television commercials is completely synthesized.

Does it matter? Certainly it does if you've devoted a lifelong career to playing in orchestras, and suddenly nobody wants to hire orchestras anymore. It matters to a lot of other people, too, even if they don't depend on playing in orchestras to make a living. Music provides only a small fraction of my income, but a large and important part of what I depend on to keep myself sane. I know of no other experience that provides quite the same kind of high as playing a really good part with a really good symphony orchestra.

Yet I also know that symphony orchestras are very expensive to run. The real question about their survival is whether enough audiences will continue to think that a live concert by a hundred living and cooperating human beings is special enough to warrant that expense. For my own personal reasons, I hope so. But I also recognize that it may not turn out that way, and someday I may have to find other ways to fill the void.

Recorded music poses a different kind of problem. Here the perceived enemy is disk jockeys, in whatever form they may take. Live musicians still make the recordings, but one recording session to make a disk that can then be played at hundreds of parties obviously provides far fewer jobs than would hiring a live band for each of those parties. Royalty payment requirements make some attempt to fill the void (did you know that part of the price of every recording you buy goes into a fund to help finance live performances?), but they make a relatively small dent in it.

Musicians understandably find this disturbing, but what can they do about it? There is no innate inalienable right to make a living as a performing musician (or in any other particular way that might strike someone's fancy). The jobs will be there as long as, and only as long as, customers find the service worth paying for. So the best, even if most difficult, course open to musicians who want to keep working is to convince the public that what they provide is worth buying.

On the other hand, sometimes it's hard to see what the long-range consequences of an innovation will be. It's quite obvious that the *immediate* effect of a growing reliance on canned music is fewer jobs, but not all effects are immediate.

When magnetic tape recording became widespread in the late

1940s, it made record production so much easier and cheaper that many more people got into the field. As the *Encyclopædia Britannica* put it, "Anyone with a good recorder and a microphone could become a record producer. Small companies sprang up in areas of music ignored by the giants: the esoteric and the avant-garde, the music of the periods before and after the highly popular Romantic classics of the nineteenth century. Chamber music, as well as Baroque works of the eighteenth century and earlier . . . flooded record stores and resulted in an unprecedented Baroque revival . . . All-Vivaldi concerts were sold out, and Bach became a best-seller."

In other words, the explosive growth of recording not only created employment opportunities for musicians to make recordings (even if the later playing of those recordings might eliminate some gigs), but it also *created new markets for live music* by awakening an interest in kinds of music the public had previously had no chance to hear. People who discovered Telemann or Mahler for the first time on records wanted to hear it in concert—so musicians were hired to play it.

Admittedly the difference in sound quality between recordings and live performances was a lot bigger then than it is now; but it's still there, and sound quality is not the only factor involved. Many people still find enough special excitement in hearing and seeing a performance in a real hall and in real time, by real people, to justify shelling out the added expense. The challenge, if musicians want to keep being hired to provide that excitement, is twofold: to make sure the public has the experience often enough to be aware of the difference, and to tame the economics so that they can afford to do it reasonably often.

And, of course, many people will still want to *make* music. The difference between listening to a record, listening to a concert, and playing a concert is something like the difference between looking at a picture of the Alps, riding through the Alps on a bus, and climbing an Alp with your own muscle and lungs. The extra satisfaction repays the extra effort so richly that some people consider it unquestionably worthwhile. In music, there are a lot of ways to get that level of personal involvement, such as singing, composing, or playing any of a wide variety of instruments. Choices must be made; any one of those pursuits involves so much learning that no one person is likely to master very many of them.

How many people make what choices will certainly affect the future shape of music, at both amateur and professional levels. You can't have many symphony orchestras, for example, unless relatively large numbers of people learn to play bowed string instruments—which has long been a major problem confronting people who want to make sure symphony orchestras survive.

It's hard to doubt that electronic instruments will attract more and more would-be music-makers. They've already been doing it for quite a few years (if you could walk into a typical music store now and then visit one from the fifties you'd be astonished at the difference in inventory), and they'll go right on doing it. An electronic keyboard or synthesizer allows a single player to do so much more of some kinds of things than any acoustic instrument that it would be unrealistic to expect people *not* to be drawn to them.

Yet some will continue to be drawn to the others, too, just as people continued to play clarinets and guitars even though pipe organs were available. A common argument for this *now* is that electronic instruments *sound* electronic. A "trumpet" or "violin" or "oboe" stop on a keyboard seldom sounds enough like a trumpet or violin or oboe to fool anybody familiar with the real thing. But this, too, will pass. It's the "primitive machine" effect, a consequence of the relatively primitive state of the contemporary art rather than an intrinsic limitation of electronics. Electronic imitations are getting better all the time, and their inadequacy will be less and less of a reason to play (or hire) an acoustic instrument rather than electronic.

I think the more effective reason will be the difference in *feel.* A synthesizer in skilled hands can give the player the exhilaration of producing complex and impressive sounds all by himself. Playing in an orchestra gives a different but comparable kind of exhilaration: that of being part of a single huge instrument whose components are human beings.

Upstart instruments like synthesizers, tape recorders, and computers broaden the range of options—just as pianos and saxophones did when they came along. They give people more choices. The future of music will be a different—and ever-changing—combination of ingredients from a growing palette. I wouldn't venture to guess exactly what it will look like at any time, but I don't think it's going to be dull. I can even imagine some peculiar hybrids, such as electronic imitations being used to augment string sections for people who want to keep symphony orchestras alive but can't find enough violinists or violists.*

A while back, in talking about the fact that even professions and available job choices evolve, I made a passing remark about "the death of the buggy-whip industry." An attentive reader named Margaret Gardiner wrote in to tell me, "Actually the buggy-whip business is doing pretty well. Connecticut alone has three manufacturers." So the buggy-whip business is not really dead—but it certainly isn't what it used to be. Connecticut also has more than three *million* residents. At the turn of the century, one buggy-whip maker per million people would have been woefully inadequate. Now it's enough, because while there are still a few people who want buggy whips, there are a great many more who want other things—many of which did not exist at the turn of the century, such as minivans, calculators, television sets, and microwave ovens.

So has it been, and so will it be, with music. There are still a few people who play harpsichords, shawms, and sackbuts—but there aren't nearly as many job openings for them as there once were. On the other hand, there are a lot *more* jobs for pianists, saxophonists, and tubaists. A hundred years from now, there will very probably be fewer openings for *them*—but more for things we haven't even thought of yet.

---

*Ironically, the day after printing this out I went to a performance of a musical in which the pit orchestra was reduced to seven pieces by using two keyboards in lieu of a string section. This is the dark side of all this: using a cheap imitation to save a few bucks when the real thing is readily available!

Taking Chances:
Risk Assessment, Philosophy, and Progress

Fear Pollution

When a country has long cultivated the habit of designating national birds, national anthems, national pastimes, national this and national that, I suppose a national neurosis is just an inevitable extension of that tendency. There may well be *several* national neuroses, but there's one in particular that I've been very much aware of recently—and from the response I get when I mention it, I gather many others have, too.

What started me thinking along these lines was a huge front-page story on a recent Sunday edition of my local newspaper, under the title "HAZARDS OF HOME." A blurb above the title read: "Home pollution inspections: Wave of the Future?" The article itself began, "The home, once considered a haven from the world's woes, is fast becoming a minefield of toxic hazards . . ."

My first reaction, after quickly scanning the boxed chart of fashionable household pollutants such as asbestos, formaldehyde, assorted water contaminants, lead, and radon, was to toss the paper aside, disgusted that it had failed to mention one of the most important pollutants currently clogging our environment: fear. I didn't actually do so; I do recognize the value of a healthy regard for real dangers, and I know that everything on that list can be one. So I read the article, to see whether it contained either information I didn't already know about them, or anything that seemed to warrant editorial comment.

I found far more of the latter than the former.

The major point of the article was a prediction that widespread fear of the items listed (and presumably others, as fast as people can think of them) will soon make testing for a wide range of contaminants a routine part of buying and selling houses. That in turn will raise the cost of a standard house inspection from hundreds to thousands of dollars, and add weeks to the already exhausting process of buying or selling a house. My overwhelming impression after reading the whole article was that the prediction is probably correct—but not because it deserves to be. I see the whole business as yet another manifestation of the recent deluge of "fear pollution," the rampant proliferation of Things We Are Supposed to Worry About. (And, no doubt quite incidentally, people who make a living trying to allay those worries.)

Other examples abound. A trace of poison found in two (count 'em, two!) grapes led to a nationwide panic that kept millions of people away from fresh fruit for weeks. Articles in newspapers and magazines advise people never to eat eggs with the slightest trace of runniness about them, because there's a slight chance that they may get sick. Many people won't eat eggs at all because they're terrified of cholesterol. Cholesterol is, of course, only one of many fashionable food worries; if that one doesn't appeal to you, or you want to cover all the bases, there are plenty of other ingredients you can find someone eager to warn you against. Eating has become, as one columnist recently observed, more like taking medicine than enjoying good food. Many people have decided, commendably enough, that smoking is not for them—but they don't stop there. They hoist militant banners and sally forth to create "a smokeless society by the year 2000," lest they be exposed to the faintest trace of someone else's smoke. A new tick-borne ailment, Lyme disease, becomes fairly common in the northeast and some people sensibly learn to recognize it and take routine precautions against it—while others swear off all outdoor activity and moan about how dreadful the world has become. One space shuttle explodes and the entire space program of a nation once famous for its courage and initiative grinds to a halt for a couple of years.

And a disgruntled editor cries out in the wilderness: C'mon, folks—enough is enough! Let's try to get at least a little sense of perspective back into things, hey?

It first occurred to me a good many years ago that worrying about all the Things We're Supposed to Worry About has done far more harm than the Things We're Supposed to Worry About themselves. I had no trouble at all finding at least one psychologist who agreed with me, and time has done nothing to change my mind. I repeat: I'm all for being alert to real, significant dangers and taking reasonable measures to alleviate them—but not making an obsession with them the central focus of everything. Collecting possible dangers, magnifying them, and worrying about them seems to have become a very serious contender for replacing baseball as "the national pastime." For now, I'll settle for calling it "the national neurosis." But I won't back down from that—and I will suggest that the national neurosis is a problem that needs treatment at least as much as most of the others.

If "the home is becoming a minefield of toxic hazards," the change is less in the home than in the occupants' attitudes. True, some homes, such as those built next door to toxic waste dumps, really do face some new and unnecessary hazards. But most of these things have been around all along. Radon, for example, was not suddenly invented in this decade. People have been eating soft-boiled eggs, pork, and assorted cholesterol sources for a *long* time, and many of them have lived to ripe old ages. In fact, more of them have been living to riper old ages recently than ever before.

We hear a great deal about the dreadful incidence of cancer and heart disease. Seldom does anybody mention that a major reason for this apparent increase is the dramatic increase in lifespan over the last century or so. Cancer and heart disease develop slowly; most people used to die quite early of other causes, so few had time to develop these. Now so many other causes of death have been tamed that more people are living long enough to get the late-acting ones, so a larger percentage of deaths is caused by those. If you look at the *overall* picture, it's much *better*, not worse, than it was for our ancestors.

The real problem is that we've been spoiled, conditioned by unprecedented medical advances into thinking we can live almost forever. People with that attitude, like many of those in science fiction stories about technologically-achieved immortality, are so determined to live as long as possible that they become pathologically afraid of every imaginable risk. People who take food like medicine, even though they have no particular medical need to do so, and spend thousands of dollars having their houses inspected for every conceivable pollutant, may live a few years longer than those who don't. But they may also realize, sometime near the ends of those long lives, that they've sold so much potential *fun* for the added time that it hasn't been worth it. In the end, eighty years spent looking over your shoulder and avoiding every pleasure that might contain some risk may be worse, not better, than seventy-five years of facing the world squarely and savoring life in big bites.

But suppose the trade-off is *not* just a few years versus more or less caution. What if you really *could* live forever?

That ability, or at least a fair approximation of it, is looking more and more like a real possibility, for at least a few people, in the not too distant future. If molecular cell repair machines like those foreseen by K. Eric Drexler and others become a workable option, people who live carefully may actually have the chance to live many centuries instead of one or less. Some researchers consider that a possibility within the natural lifetimes of people already born, and some people are already trying to improve their odds by having themselves cryonically suspended upon what is now considered "clinical death." People with access to such nanotechnology will have much more at stake than a few years more or less of more or less pleasant life. How will that change their attitudes toward routine risk-taking?

If cell repair nanomachines become good enough, of course, the definition of careful living may change, too. Machines that can repair virtually any damage at the cellular level may consider cancers and heart muscle or artery deterioration just another routine maintenance problem. If that's the case, people may go to the opposite extreme from the one I'm now grumbling about. It may become fashionable to eat, drink, and be merry with little concern for any danger short of sitting on an atomic bomb.

But it's easily conceivable that at least the first cell repair machines may be a bit more limited in their capabilities. Maybe they'll be able to repair relatively small deviations from normal physiology, but only if the body's owner makes some effort to maintain a basically close-to-optimum operating environment. Suppose, for the sake of argument, that you have access to nanotechnology which can let you live for a thousand years, if and only if throughout that time you take care of yourself in a way approximating what many people are advocating in 1989. Is it worth it?

We've all read those stories about future societies with more or less immortal inhabitants, stagnating because their inhabitants placed such a high value on surviving every possible minute that they were afraid to risk *doing* anything. We haven't quite reached that point; we don't yet have technology that lets most people believe seriously that they are likely to live much more than a century. We do have technology that lets them believe they can live a few more years if they're very, very careful than if they're not. When I see how many of them react to that option, and extrapolate the trend, I can easily imagine that the familiar picture of cowardly stagnation is eminently plausible. If people are going to react that way to the possibility of even slightly longer life, the chance for *much* longer life may pose a very real, and large, psychological threat.

I think that possibility warrants concern and consideration—but not resignation. Remember, we're new at this. It's only quite recently that human beings have been able to consider even present life expectancies a realistically achievable goal. They haven't had much time to think through, and sort out their feelings about, how such options should affect their fundamental outlook on living. It may be that, with more time, people will work those things through to saner philosophies than any of us have yet imagined for living with their new potentials.

In the meantime, though, as I look around me at what people have done with their opportunities so far, I can only hope that the attitudes I see are a passing phase and people *will* grow beyond them. I see people digging diligently to find things to worry about, and taking every precaution anyone suggests might conceivably gain them an extra minute of life—and I wonder why many of them *want* every possible minute of life. They don't seem to enjoy the ones they already have; *joie de vivre* seems foreign to their natures. I can easily understand taking reasonable precautions to live a long time because you love life so much you want all you can get. But it seems to me that a great many people are now taking every possible precaution not because they love life, but because they fear it slightly less than they fear death.

But only slightly. And that seems to me a terribly sad way to live.

Foolproof

In 1990 *Analog* published a story called "Funnel Hawk," by Tom Ligon, in which the heroine flew a specially designed airplane quite close—and on one occasion into—tornadoes. One reader objected vehemently to the story on the grounds that it might lead irresponsible young pilots to try to repeat the actions it described, with disastrous results. If even one young person died as a result of that story, this reader suggested, the author and I would forever bear the guilt for his death; and she hoped we would never again do anything so reckless.

I sympathized with her concern and gave her argument careful consideration, but in the end could not find it valid. No one dies as a result of reading a story. If the scenario the reader feared actually occurred—which I consider extremely unlikely—the death would result not from Mr. Ligon's well-crafted and widely admired tale, but from the pilot's own stupidity or irresponsibility.

The story made it very clear that the plane it featured was no ordinary craft, and the woman who flew it was no ordinary pilot. The plane had a multitude of special design features enabling it to survive and maneuver under conditions most planes couldn't handle. The pilot was not only very talented but very experienced—and she *still* had trouble, and knew what she was doing was risky. Anyone who attempted those tricks without a comparably special plane and training would be either an idiot or a fool.

Idiots don't fly, at least legally; the requirements for earning a license are much too demanding. Fools *may* fly, on occasion; flight school cannot guarantee that its graduates will always exercise good judgment. But if one fails to, to what extent is anyone else responsible for his actions?

Hardly any, as I see it. Indeed, it seems to me that many of our culture's recent problems stem from a failure to expect people to accept responsibility for their own actions. By itself, I would not consider that letter sufficient subject for an editorial. But as only one of many manifestations of what seems to me a widespread trend, it's a good starting point.

The current fashion is to place the blame for foolish behavior anywhere but where it belongs: on the fool himself. One result of that is that people who should be facing the fact that *they* need to shape up are instead encouraged to blame (and sue) other people who in some distant way helped make their foolish behavior possible. Another is that manufacturers are expected to build everything so that any idiot can use it, and writers and editors are expected to avoid even mentioning anything that might give somebody the idea of trying something stupid.

Now, I'm all in favor of building things so that they can be used with a minimum of unnecessary difficulty and are not likely to malfunction dangerously if used with reasonable prudence and skill. I'm writing this on a computer that is far easier to use than any of the several I used before it, and I consider that progress with a capital P. I see no advantage at all in making something that's hard to use when you know how to make it easy and still capable of doing good work. Furthermore, I wouldn't consider it responsible behavior for a writer or publisher to publish "how-to" advice which he knows is wrong and dangerous.

However . . .

Fiction was never intended to be read as an instruction manual. And I see a big, crucial distinction between designing a thing for optimum performance when responsibly used, and insisting that everything be so designed that it can be subjected to any amount of irresponsible abuse without adverse consequences. The obligation to make things "foolproof" does not extend so far as making them absolutely safe for even an absolute, genuine, literal fool.

Some things *cannot* be made so that any idiot can use them. Airplanes, for instance. Flying one is a complicated business, depending on an integrated combination of highly specialized skills that can only be developed by training and experience. Using those skills also requires a finely developed sense of judgment. An airplane in the hands of somebody lacking those qualifications is, without a doubt, a lethal weapon. This does *not* mean that airplanes should be banned, or that a moratorium should be declared on manned flight until airplanes *can* be made so that any unskilled dolt can fly one safely. It does not even mean that writers and publishers should refrain from writing about pilots pushing their skills beyond the point of absolute safety. It *does* mean that you take effective measures to keep people who lack the requisite skills and judgment from having command of aircraft.

An airplane is only one example of a tool which is highly valuable in skilled hands but cannot be used by just anyone. Some of the others are not so obviously dangerous, but are just as dependent on special skills. Musical instruments, for example. A Stradivarius violin can sound good enough to make thousands of people spend dozens of dollars each to listen to it. It can also be an instrument of torture in the hands of a player who hasn't spent years learning what to do with it.

Nor is the phenomenon limited to *complicated* technology. People can find a way to make practically *anything* malfunction, or to turn it from constructive to destructive uses. As simple a tool as an axe or a knife is surely a dangerous weapon when wielded by someone who chooses to treat it as such, or who hasn't learned how to handle it properly. It's also an extremely *useful* device in the hands of someone who understands the benefits it can produce, and how to protect himself and others from the incidental dangers.

Completely aside from the fact that we *need* the beneficial effects of many tools which can't be made completely foolproof, a case could be made that there's some intrinsic value in having a few around that can be dangerous if misused or abused. Our culture has tended to remove as many evolutionary pressures as it could from our species, by making it easy for just about anybody to survive. There are advantages to this, of course, both in terms of humanitarian values and in terms of actually contributing something to the species' future. Some of our most productive minds have been in bodies which would not have lived long in centuries or even decades past. This has been to everyone's advantage and credit. Even in cases of mental disability, most of us would agree that today's more compassionate attitudes are an improvement over those prevalent in the past.

But the fact that society here and now can support and encourage members who would not have survived in the past does not result entirely from our being persons of better will than our ancestors. That is, in fact, questionable. Our more "enlightened" behavior is, to a greater extent than most of us would like to admit, a luxury which we can afford because of modern medicine, agriculture, and other technologies. The technological infrastructure that makes it possible is big and powerful, but also fragile and vulnerable.

Civilizations *can* collapse, as history has shown repeatedly. If ours should do so, the next is likely to arise sooner and more successfully if the survivors include a good number of people who are well prepared to cope with a wide range of challenges. They would need, in other words, intelligence, competence, and good judgment. Tools that require a certain amount of those qualities are one of the few things we have left that tend to select for them. I'm *not* suggesting that axes and knives should be left lying around among people who have been diagnosed and isolated as unable to function as parts of normal society. But among those who are *attempting* to function as parts of normal society, I see no reason why specialized skills and common sense should not be among the things expected of them—or why lapses in those areas should not carry their own penalties.

Such arguments, of course, make many people extremely uncomfortable (and hot under the collar). If you don't like them, feel free to skip over them—but please consider the next. I see one more danger in the current fashion for making everything Absolutely Safe, and that is related to another kind of evolution: cultural.

A while back I mentioned the concept of a moratorium on manned flight until it can be made perfectly safe. I haven't actually heard anybody suggest that—but I have heard people propose something very similar in a variety of other fields, including spaceflight, genetic medicine, and the various avenues of research leading toward nanotechnology. Not doing things until they're perfectly safe would be nice, but it's not realistic. There's a lot of truth in the old saw that you have to walk before you can run, and you have to crawl before you can walk. What we seem to be forgetting is that a baby making the transition from crawling to walking *always* takes some falls. One who was afraid to stand up because he couldn't face the prospect of falling would *never* learn to walk, much less run.

Our civilization, it seems to me, is in a position rather like a baby who has suddenly realized that he's standing up, and is frightened by the implications. It's a special case of a little knowledge being a dangerous thing. We've learned to do a lot of things that our ancestors could barely dream of; every one of us, every day, does many things which could have gotten us burned as witches three hundred years ago. We've learned to do some of them so well that we can do them with a very high level of safety and reliability.

And we've become spoiled. We've come to think that since we can do *some* things so well and safely, we *shouldn't* do anything that we can't do equally well and safely. And therein lies a terrible pitfall. You *cannot* demand the same standards of safety and reliability from a new technology that you do from a mature one. Major advances in our abilities have *always* involved doing things that were later recognized as dangerous, and doing things ineptly as the first step toward learning to do them expertly. Our ancestors seem to have understood this better than many of us. Perhaps because they hadn't learned to do anything as reliably as we now do a few things, they took tragedy in stride as a normal part of exploring frontiers.

G. Harry Stine spoke of this in an "Alternate View" column called "Where Have All the Heroes Gone?" (December 1990). At its end he quoted M. Stuart Millar, owner of the Piper Aircraft Corporation—who subsequently wrote Harry a letter commenting on that column and containing another memorable quote: "No one believed we should cease Antarctic exploration because Captain Scott's group died in the effort. No one believed we should cease efforts to fly from New York to Paris because Captain Nungusser disappeared while trying. No one thought we should cease attempting to fly to Hawaii because crews were lost seeking the Dole prize. No one thought Pan American should cease Pacific flights because Captain Music's Philippine Clipper disappeared on a survey flight. That is the price of the uniqueness of the human being compared to all other terrestrial animals."

Our civilization—which shut down an entire space program for a couple of years because of one accident—badly needs to relearn that fact. *No* technology—indeed, no human activity—is absolutely safe; and new ones are inherently cruder and more dangerous than old, highly developed ones. If we allow the aggressively timid to demand that new ones meet the same standards as old ones, we can forget about playing an important role in the great adventures and opportunities for better living that lie ahead.

They're there; in space, and nanotechnology, and a host of other fields that beckon from the horizon. But getting there will involve, as it always has, risks—sometimes even unto death. Rather than using that as an excuse to huddle in the corner and fear what's outside, we should be grateful that there are those willing to take those risks—and be supportive of their efforts.

For that will always be the way to the greatest rewards.

Toward More Perfect Governments, Big and Small

Experience Required

I'm not sure when or where I saw the cartoon, except that it was on an editorial page of my local newspaper. So I can't quote it to you exactly, or credit the cartoonist by name. What I can do is give you the gist of it, and I think you'll appreciate the point it made.

The first of its four panels, as I recall, showed one character representing the federal government telling another, "Our expenses are bigger than our income. What can we do?"

"No problem," says his colleague. "We'll increase what we take from the states and decrease what we give back to them."

The second panel showed one character representing a state government telling another, "Our expenses are bigger than our income. What can we do?"

"No problem," says his colleague. "We'll increase what we take from the towns and counties and decrease what we give back to them."

The third panel showed one character representing a local government telling another, "Our expenses are bigger than our income. What can we do?"

"No problem," says his colleague. "We'll simply increase the tax rates we charge our citizens."

The fourth and last panel showed a taxpayer standing alone and forlorn, with his empty pockets turned inside out, saying, "Tell me about it!"

Meanwhile, an article in the same paper, close to the same time, described the ongoing efforts of legislators to trim a state budget—while keeping for their own members such perks as multiple luxury cars and luxurious residences "rented" from the state at rates ridiculously lower than the expense of maintaining them.

It's almost enough to make mere mortals want to drop whatever they're doing and go into politics. Governments, after all, have an enormous advantage over everybody else; and the bigger they are, the more pronounced that advantage is. When a government decides it needs more money, all it has to do is announce that it's going to take a bigger percentage of the income of somebody else with less clout—and then take it, by force or at least the implied threat of force.

The individual taxpayer has no such recourse. If his or her expenses are getting beyond his means and he is conscientious enough to be unwilling to sink into debt or go on welfare, he has two choices. He can increase his income, by doing or making something that people are willing to pay more for. Or he can find ways to live more frugally, like doing without some things he might like to have, shopping more carefully, and taking care of things so they'll last longer.

Governments are the *only* entities which can demand more money *whether or not* the donors are willing to pay it, or believe that they're getting their money's worth. Furthermore, governments gloss over the exceedingly important distinction between an increase in an absolute quantity of money and an increase in a *percentage* of an absolute quantity. *Of course* governments need more money to get by each year. We all do, because inflation we have always with us, to greater or lesser extents. But for most of us, in more or less "normal" times, that increase in need is at least approximately offset by increases in income. Sure, I pay far more for things now than I did twenty years ago; I also earn far more than I did then. The same is probably true for you.

Yes, I know there are exceptions. I've been one of them; I've had spells when my income was not increasing nearly as fast as general inflation, so that for a while my real purchasing power was declining. Over the long term, though, and over the population as a whole, both income and expenditures tend to rise. Since most taxes are based one way or another on percentages of what people earn or spend, this means that governments *automatically* get more tax revenues—by amounts roughly commensurate with their increased costs—*without* changing the tax rates expressed as percentages.

When they increase the percentage rates, this gives them a double whammy: they take an increased percentage of an increased amount! And since ultimately that extra increase can come only from the taxpayer's pocket, it tends to offset the increase which allowed him to compensate for inflation. The result is that the individual taxpayer, more than anyone else, finds it harder than he should to keep up with inflation—and the peculiar way taxes are increased is directly responsible.

Say inflation is 5 percent. If you get a 5 percent raise *and keep the same fraction of it you always did*, you're effectively back where you started. You haven't made any real gain, but neither have you lost any ground.

Now suppose that before the inflation and raise you were paying 20 percent of your income in taxes. If that rate doesn't change, the government gets the customary 20 percent of your increased income. Both you and the government still break even; you both still have approximately the same purchasing power you did last year.

But if the government increases your tax *rate* from 20 to 21 percent, that means that an additional 1 percent of your total, increased income now goes to the government. Now your real, spendable income has increased by *less* than the cost of living, while the government's has increased by *more*. You have most assuredly lost ground, while the government has gained substantially more than it should have needed to keep up with inflation in the same way that you're expected to.

So *you* have to find ways to live more frugally (or to make people *want* to pay you more), while governments seem (despite periodic bursts of lip service to the idea) chronically unwilling or unable to do either. (I respectfully submit that allowing legislatures to keep fancy cars and houses at taxpayer expense does not bespeak serious frugality.)

I suggest that the problem is twofold:

1. **Governments have little incentive to spend frugally, or to provide services that people value so highly they *want* them even if they cost more, simply because they can get away with not doing so.** They have the power and immunity to take pretty much what they want, and they don't have to deal with competition for the services they provide.
2. **They may literally *not know how* to do more with less, because they've never had to learn.** This may seem a flippant suggestion, but I offer it quite seriously. Again, I acknowledge that there are individual exceptions; but quite often legislators come from well-to-do families and therefore have stood, so to speak, on the shoulders of financial giants. The skills required to live on an inherited fortune are not at all the same as those required to live on an income barely adequate for subsistence.

Changing any of this, of course, would require changing laws. The biggest problem in that is that laws are made by legislators, and it would hardly be surprising for them to prefer things as they are. Nevertheless, just for fun, what changes might be desirable if we could figure out how to make them happen? What might we do differently if we were designing a system from scratch?

I don't plan to say much about Problem (1). Figuring out how to make legislators more accountable (elections help, but not enough) is just a bit more than I feel ready to tackle this morning. The same goes for trying to provide competition, or some other incentive, so that governments have to think harder about the *quality* of the services they provide for the money they collect. (The same legislature that was keeping its luxury cars and houses proposed deep slashes in funding for frills like education and road maintenance!)

However, I do have a simple, concrete change to propose for Problem (2). I have suggested that the problem is that many legislators lack experience in making good use of limited funds. Therefore, let us require them to have some of that experience, as one of the qualifications for the job. Let's require our lawmakers to have spent a certain number of years living in something at least approximating poverty, without going on welfare or getting in trouble with the law! That, I suggest, would constitute good evidence of ability to use limited resources wisely.

The requirement might be waived, of course, for somebody who started out in poverty and managed to rise above it, through his own efforts and without illegal activities, in less than the stipulated time. That would be evidence of another talent that governments urgently need: the ability to improve one's own income without recourse to forcibly taking it from others less powerful.

What we *don't* need more of is politicians who have never had to learn to get the most out of limited funds, but are all too adept at simply dipping deeper into the pocket of someone else unable to resist.

The All or Nothing Fallacy

A character of mine once observed that rationality, when it exists at all, usually exists only as a transient state on the way from one extreme to another. Part of the problem, I think, is that extremes—beliefs that *everything* must be done thus, or *nothing* should be—are much *simpler* than the middle ground. If you can sincerely believe that everything is black or white, or that eating chocolate is always good or always bad, your life is (or at least seems) much simpler than if you have to try to judge just how gray things are, or how much chocolate might be how good or bad for which individual. It takes so much less *thought* that way, and thinking is *work*.

Of course, reality isn't always so obliging. It has this annoying habit of actually coming in many shades of gray (and even colors!). Metabolisms vary from individual to individual, so that a chocolate habit which brings one person a lot of pleasure and few, if any, bad side effects might be very bad, perhaps even lethal, for another. So simplistic beliefs that everything or everybody has to be treated the same way, whatever that way might be, tend to lead to inappropriate decisions and undesirable consequences.

When that happens often enough, people begin to notice and decide that the simplistic creed they've been following isn't so hot after all. So they throw it out—baby, bath water, and all—and start doing just the opposite.

Until they realize *that* doesn't always work, either. So, having had time to forget, they start moving back. . . .

So it is that much of history is the story of cultures oscillating between philosophical extremes. It is valuable to have consistently applied principles, provided they're well formulated, take into account all relevant facts, and don't get distorted by facts that are *not* relevant. Unfortunately, that isn't as easy as it sounds—if only because real situations tend to be well cluttered with facts, and it's not always obvious which ones *are* relevant.

Consider, for example, the question of personal freedom. In many times and places, society—through government, church, and/or custom—dictated practically every move in everybody's life. Such conditions have led, not surprisingly, to reactions in which it became fashionable to think that individual freedom is paramount and people should be free to "do their own thing." Both extremes, of course, have their problems, and tend to lead to reactions going back toward the other extreme. Hardly anybody ever seems to think of trying to stop at a sensible position somewhere in between.

Many educators and employers, for instance, have strictly regimented every aspect of their students' and employees' behavior at school or work, imposing, for example, strict, detailed dress codes. Others, seeing no real relevance of such details to the job at hand, have been much more laid-back about them—but sometimes an easygoing attitude about peripheral details extends to a general laxness about everything, including the work itself. People living under either "all or nothing" philosophy—strictness about everything, or strictness about nothing—tend to find enough cause for dissatisfaction to drive them toward the other extreme. A lucky few manage to stop, at least for a little while, before they get there.

Clearly, if a society is to function well, the people doing its work must be held accountable for doing it well. A general "anything goes" attitude simply won't work and can't be tolerated. It *does* matter, very much, that a pilot or a surgeon take pains to do things right, all the way down to the smallest detail. It even matters that a grocery stock boy shelve merchandise in the right part of the store. If many of them fail to do so, customers won't be able to find what they need and the store will lose them and go out of business.

But it doesn't really matter what color shoes any of them wear—pilot, surgeon, or stock boy. Many employers and educators have demonstrated that it's quite possible to demand, and get, adherence to high standards in the areas that actually matter, while remaining quite flexible about peripheral things that don't.

Of course, what things matter varies from job to job. As a writer, I *must* meet deadlines, but my publisher doesn't care what kind of weird hours I might spend at the keyboard to do it. (I've known several hundred professional writers and have yet to meet any two who work the same way.) As a musician, I must be onstage and ready to play at the scheduled time for a rehearsal or performance, every time. In one case, being in a particular place at a precisely specified time is essential; in the other, it's irrelevant. Again, it's a case of judging what's appropriate for a particular situation rather than categorically assuming that all situations must be treated the same way. (When I'm being a musician, even the color of my shoes can matter! The appearance of an ensemble is part of the show, and scuffed white loafers on one member of an orchestra dressed all in shiny black can definitely distract an audience.)

Judging what is actually important in each situation, so you can concentrate on that and "not sweat the little stuff," is, of course, a challenge. Some institutions and individuals—military organizations are the classic example—feel that it's so important that some of their jobs be done exactly right that they dare not risk leaving *anything* to individual judgment. Military people put up with it, because they have no choice and/or view their service as temporary and important, but many civilians won't. Thus dictating how everything is to be done generates one set of problems, and the "obvious solution," abandoning all standards, generates another. Both are genuinely problems, and neither is genuinely a solution.

The tendency to swing between simplistic extremes tends not to affect just a few particular lines of work, but to permeate a culture's entire mindset. This country, in the 1950s, placed a great deal of emphasis on conformity: on "belonging," "fitting in," and generally doing What Was Expected in everything from raising the right kind of family, to going to church on Sunday, to wearing the right kind of socks. Enough people eventually got tired of that to produce a reaction: such an extreme emphasis on individual freedom in the sixties and seventies that some people, lacking either a strong personal philosophy or strong social guidance, drifted with no clear idea of how they should (or even wanted to) live.

Now (quite predictably) we're seeing a reaction to *that*, in a growing call for more standards and enforcement. Extreme examples recently in the news as I write this include a widespread clamor for school uniforms and curfews for everybody under a certain age. Imagine the outrage if somebody suggested imposing similar requirements for all members of a sex, a race, or a religion. (Remember when—and where—Jews were required to wear stars?) Treating children differently from adults is not, of course, quite the same thing; but the difference is not as absolute as all that. Nor is "childhood" a state defined with absolute sharpness and clarity. Even if you accept (as most societies do) that children *must* be treated differently in some respects, that does not automatically justify any and all differences. Nor does it eliminate the need to think carefully about where, and how sharply, the boundaries are to be drawn.

The simple fact is that such movements as "uniforms and curfews for all" can be seriously considered only because the target group doesn't have the clout to defend itself. They have the serious defect of deciding how to treat people on the basis of belonging to some demographic group—which is as much Applied Prejudice as if the target group *were* women or men or blacks or whites. It may save the trouble of learning which individuals deserve such treatment; but it punishes the innocent along with the guilty, hardly an ideal of the best periods of American history.

Let me propose a better way of thinking. Instead of saying that all people under a certain age must be off the streets before a certain hour (the confusion of equality with fairness), let's take the trouble to look closely at which ones are actually causing trouble. Then let's deal firmly and in no uncertain terms with *them*—instead of making excuses for them and letting them off with no consequences for their misdeeds. Instead of saying that *all* actions must be left to individual whim, or that all must be dictated, let's take the trouble to recognize that *some* actions cannot be tolerated, and then be consistent about not tolerating them—while not meddling with those that do no harm.

In other words, let's use our brains and some rational discrimination and gumption to make a way of life which is actually better, and lets people of responsibility be treated as such—while making them *earn* that right. That will, of course, require large numbers of people to learn to think clearly enough to damp down those oscillations between "all" and "nothing." But wouldn't it be worth it? And are you *sure* it can't happen—eventually?

Unlicensed Practice

If your car's brakes were failing, who would you want to diagnose and repair them: your favorite mechanic, or his bookkeeper?

If you answered "bookkeeper," please remind me not to ride with you. Like many people who've thought about it, I have a rather firm conviction that important decisions requiring specialized expertise should be made by people who *have* that specialized expertise. I know this opinion is widely shared, and often formally recognized by the law. Practicing medicine without a license, for example, is a criminal offense in every jurisdiction I know about.

At least, for individuals.

Corporations, it's becoming increasingly evident, may be a quite different kettle of fish. If I asked you a question patterned after my opening line—"Who would you rather have decide what to do about your chest pains: your doctor or your insurance agent?"—I'm reasonably sure that you would answer, promptly and emphatically, "My doctor." Yet in the maze that presently passes for a health care system in this country, it's growing ever more likely that the most important decisions will be made by your insurance company.

Some examples:

- An orthopedist recommended magnetic resonance imaging to help diagnose a troublesome back problem. The patient's insurance company refused to pay for it.
- A young, active woman broke a foot. Her insurance covered treatment for the fracture, but when the doctor declared it "healed," she could still only walk with a pronounced limp, at a top speed about half normal—a situation clearly far short

of a complete cure. Her doctor prescribed physical therapy. The insurance company said they wouldn't pay for it because it was "medically unnecessary" since she could walk across a room by herself. (On appeal, it reversed the decision and did agree to pay—but only under protest.)

- Lest you think these mere isolated cases, this part of the country recently had a highly publicized controversy over the express, blanket policy of certain health insurers of paying for only a single day of hospital care for new mothers and their babies—regardless of how much their doctors thought they needed.
- At a briefing intended to help employees decide which of several medical insurance options to choose, I posed the following hypothetical question to a spokesman for one plan. "Suppose I have already been seeing a specialist for an irregularly recurring condition, and am under standing instructions from him to come in once a year for a check-up. This specialist is on your approved list. If I switch to your plan [in which specialist visits are supposed to happen only on referral by a 'primary care physician'], do I really have to make an extra visit to a general doctor every time I want to go for one of the specialized check-ups I already know I need?" The answer was, in essence, "We might be able to work out a 'standing referral,' depending on the circumstances of the particular case; but we [the insurance company] would certainly want to review it."

In each of these individual or collective cases, a medical decision—a kind of decision people spend years learning to make—was being made by an entity having neither that specialized training nor familiarity with the patient's medical history. You might argue that the insurance company wasn't *really* making a decision about how to treat the patient, but only about who should pay for it. If the patient is determined to have the treatment, he can save his pennies and pay for it himself.

Sure, if the ailment in question is, say, a hangnail. For more serious and complicated matters, in case you haven't noticed, various factors that would take me too far afield have made medical care so expensive that many people *can't* afford much of it on their own. Whether this is a desirable, perhaps character-building, state of affairs is a question I don't plan to get into. The important fact for now is that, rightly or wrongly, a decision about whether an insurance company will pay for a treatment is *de facto* a decision about what treatment a patient will receive—that is, a medical decision.

You might also argue that such decisions are not actually made by "companies," but by individuals or groups of individuals who work for the companies, presumably operating within strict policy guidelines. Doesn't my argument fall on its face if those decision-makers *are* doctors (who happen to be employed for that purpose by insurance companies)?

Not really. In the first place, I strongly suspect (though I admit I haven't proved) that they seldom *are* doctors. In at least one case that I have some familiarity with, at least the original decision seems to have been made routinely by a nurse. If sometimes the insurance company's arbiter *is* a doctor, that may technically take it out of the category of "practicing medicine without a license," but only technically. There's still something crucial missing from the picture. Why should a doctor who hasn't even met the patient be able to overrule one who has?

As for decision by committee, I realize that in principle a committee is supposed to make better decisions than an individual by having several individuals compensate for each other's blind spots and biases. In practice, it is not for nothing that someone once observed, "A camel is a horse that was designed by a committee."

No, folks, I've looked at all these angles and they keep leading me back to my original position: The person best qualified to make a medical decision is a competent, conscientious physician who is thoroughly familiar with the patient and his or her problem.

"Ah," you (or an insurance spokesman) may say, "there's the rub: Not all doctors are equally competent and conscientious. Some haven't kept up on their skills, and some have become too lazy to put as much thought into every case as they should. Why, some have even been known to try to bilk insurance companies! That's why we need someone to oversee them and make sure that they're really doing what the patient needs."

I'm with you all the way up to the last sentence, but there we part company. Of course not all doctors are equally good or honest. Can you think of *any* field in which all practitioners are? Yes, patients need to try to pick a doctor who is good for them, and that isn't always easy. Most of us appreciate all the help we can get—but many of us do *not* want someone else making the decision for us, even to the extent of allowing us to choose from a short list they provide.

And we certainly don't want people who know finance but not medicine making our medical choices primarily on the basis of cost. Yes, costs must be considered and controlled (and I do appreciate the insurers' need to protect themselves against doctors who pad their bills). But the patient's needs—which are a highly individual matter—should be considered *first.* The treatment given should be the *best* treatment that can be afforded. Somehow, it seems, the cart has been allowed to overtake the horse.

Part of that problem, I think, can be seen in a recent fashion among hospitals for telling their employees to think of patients not as "patients," but as "customers." There can be some value in this, if the object is to remind them that patients are paying for a service and deserve to get what they're paying for. But there's also a danger of thinking of them primarily as sources of revenue. Maybe everybody needs to be reminded that they are *both* patients and customers. They are *primarily* patients: people who need their medical problems solved. Incidentally to that, they are customers who provide revenue for the caregivers—and in that sense they are customers who deserve their money's worth.

The emphasis *may* be beginning to shift back to where it belongs. In some areas, groups of physicians have begun to form their own Health Management Organizations, with the accountants working for the doctors instead of vice versa. Dr. David Finley, the president of one such network in the New York area, was quoted thus by Gannett newspapers: "We feel we're in a better position to make decisions on patient care than someone in an office in an executive position. In a traditional HMO, the business people make the decision, and the physician has input. I'd rather see a health-care decision rest on a physician advised by business people, rather than the other way around."

Which sums it up very nicely, I think. The physician-operated networks *will* need advice from business people, because they face quite considerable financial and legal obstacles to make their venture work. But they're so clearly on the right track—trying to get the "advice" flowing in a more appropriate direction—that I, for one, very much hope they *will* make it work.

Natural Succession

A wise man once observed that companies, like ecosystems, go through fairly predictable stages of development. Lately I've been pondering the disturbing possibility that entire cultures—ours, for instance—might follow much the same pattern as companies.

A forest starting over after a fire initially has only certain kinds of small plants, which are in turn replaced by a sequence of mixtures of other types. The earliest to thrive will be types that like lots of sunshine, because there is no shade. As some of those become large trees, small understory plants that like sunshine necessarily give way to those that need shade.

Companies, according to the aforementioned observer (who has had many years of industrial experience), are typically started, and initially run, by technical or "idea" types. Later they come under the supervision of professional managers, and eventually the operation is largely run by its accountants. Each phase tends to bring with it certain characteristics of *how* the business is run.

There are exceptions, of course, but the pattern does seem to have a good deal of validity. The technical/idea phase might be exemplified by the birth of the Apple computer company, started by people with lots of hands-on experience building and using computers, who had an idea for a new kind of computer that would appeal to people put off by the old kind. Or by some of the early publishers of science fiction, who got into the business because they personally loved (and sometimes wrote) the stuff and wanted to get more of what they liked into print, and make a living in the process.

It may be inevitable that companies run by such people eventually turn over much of their management to professional managers: people whose main interest and training is management *per se* and for whom the product they're selling is incidental. Those who are driven by a passion for research or art may find the business aspects of running a company an annoying chore, and may not be very good at it. As a small company started by individuals with a Really Good Idea becomes a big company trying to meet a large demand, the Really-Good-Idea folks may find that they need help in dealing with problems like optimizing manufacturing processes and marketing strategies.

So they bring in MBAs, folks who have specialized in just that sort of thing. The potential problem with that is that professional managers may view managing as an end in itself. It's their job, and although they may try to do it conscientiously, they may have no particular interest in the particular product whose manufacture and marketing they're directing. They may not even know anything about the product, and may even sneer at the very suggestion that they should. I have actually heard some managers say, and claim that they were taught in business administration school, that a good manager can manage anything, and doesn't need to know anything about the product his or her current employer is selling.

I have long found this attitude disturbing. It seemed obvious to me that you can't make rational decisions about how to make or sell something if you don't know what it does or how it does it, how it's made or the potentials and limitations of the manufacturing methods, or what kind of people buy it and what they want to do with it. I was recently persuaded to make a *small* adjustment in my views, by a man who I believe *could* manage a good many processes without understanding their details. The key to doing so is to recognize and admit what you don't understand, and *listen* to your employees who do.

However, too often it doesn't work that way. Pride may make it difficult for a manager to admit that someone lower in the organization chart knows something important that he or she doesn't. And a manager in that position is at the mercy of his employees: He must take their word for what they tell him, since he is not qualified to judge for himself whether it makes sense or how it might be improved. It works fine *if* the employees are competent, conscientious, and trustworthy, but can cause big trouble if they're not.

So, while a skilled manager can *sometimes* effectively supervise a process he doesn't know much about, there's no real substitute for understanding at the hands-on level. Without that, even a conscientious manager may not realize until too late that he's being undone by employees who don't really know or care what they're doing. Or he may alienate employees who do, by treating them as if they don't because he can't tell the difference. Perhaps worst of all, he may become unduly reluctant to take a chance on new ideas from the Really-Good-Idea people. (Remember them?)

In its more extreme forms, this leads naturally into the third stage, where the accountants are effectively making most of the business decisions. Managers who can't personally judge whether a product will appeal to its intended market, or whether a proposed manufacturing process is feasible or efficient, may make more and more of their business decisions on the basis of what their accountants tell them. What the accountants tell them is important, of course; any company that wants to stay in business must keep a close eye on what it spends, what it gets for that, and what it can expect to make in the future.

But while accountants can provide highly specific, hard data on the first two points—direct costs and benefits—they are far less able to forecast future revenues. So the manager who wants to keep his job may make his decisions more and more on the basis of the easy data and try to avoid guesswork about the hard. This may be prudent in the short term and disastrous in the long. Companies that Make It Big are likely to do so by taking a chance on something new, and companies that look only at immediate costs and direct benefits become less and less likely to do that. Eventually they may be left behind and competed into the ground by those more willing to take a chance.

In extreme cases, they may even seriously undermine their own ability to keep doing what they already do well. I am reminded of a story I was once told, and assured was true, by a source I consider reliable. A company whose main product was jet engines was taken over by another, apparently well into the "government-by-accountants" phase, whose prior experience was in supermarkets. The new owners promptly did a profitability study of their new acquisition by dividing the profit attributed to each department by the floor space it occupied. On that basis, they abolished their two "least profitable" departments: metallurgy and fluid dynamics, the very foundations of jet engine manufacture!

What does all this have to do with whole nations or cultures? Well, I think there's evidence of similar patterns—and perhaps similar lessons to be learned—on at least two levels: formal government, and general trends in How Things Are Done.

The United States of America, for example, started (in the sense that it can be considered to have started with the Revolutionary War and the few years thereafter) with a bunch of idea people. Sure, their original motivation was political; they were fed up with the old boss and ready to strike out on their own. But they had what were then some pretty radical ideas about how to do it better, and managed to drum up considerable enthusiasm among the customers for those ideas.

As time went on, people thought less and less about the ideas. The country became increasingly wrapped up in the day-to-day problems of running itself, and that running was increasingly done by career politicians. Sure, some of them did remember the ideas and ideals, and did what they could to implement them. But more and more, the business of politics was politics, and more of most politicians' effort went into getting reelected than into doing what would be best for the country. This, pretty clearly, is the "professional manager" phase.

Are we now into the "accountant" phase? This seems to me less clear, at least in the governmental arena. Anyone who's looked at the national debt and many government spending policies might at least question whether the government is being run by *good* accountants. But at the general cultural level, I've been seeing a wide range of areas in which accountants seem to be the dominant force in making decidedly non-bookkeeping decisions.

A few examples:

- Just today I heard a radio interview with a highly regarded concert pianist, comparing current and former practices in music recording and distribution. In his earlier years as a recording artist, he said, the first priority was to make the best possible recording of the best possible performance. If a session wasn't going well, producers and performers would take a break and come back to try again later. Now the emphasis is to get the recording done and out, with musical quality a secondary consideration. Whether a particular recording is made at all is likely to be decided beforehand not by musicians, but by a marketing committee.
- Most radio and television stations decide what to do largely on the basis of how much money they expect to be able to make. New York City has a big enough population to support commercial stations with almost any kind of programming, but what they do is dictated largely by their sponsors' concern for the bottom line. Thus we have such spectacles as a station which had for years given a convincing appearance of actually caring about high-quality classical music programming, changing literally overnight and with no warning to the hardest of hard-rock formats. At least one other station has switched most of its programming to short pieces and excerpts from big ones. Full-length symphonies and concerti, performed in their entirety, have become a rarity because advertisers have convinced programmers that most listeners have short attention spans (which may or may not be true) and the few who don't, don't matter.
- I have complained on several previous occasions, as have many other people, about the fact that important medical decisions are increasingly made not by doctors, but by insurance

company clerks. A good snapshot of the currently prevalent attitude toward medicine, I think, can be found in the ads for one HMO that brags about its "commitment to involving doctors in the decision-making process for their patients' care," as if this were a remarkably magnanimous act on their part. To some of us it still seems self-evident that doctors should be the *primary* decision-makers, but our culture is being rapidly brainwashed into believing otherwise.

- When many science fiction book publishers were personally interested in the field at the literary level, there were two phenomena that have become largely extinct. The "midlist" included most books; they didn't become best-sellers, and nobody expected them to, but they did include a wide range of books appealing to a wide range of tastes. And they did make money, if slowly, as part of the "backlist," books printed some time ago but kept available and largely sold by mail order from ads in the backs of other books. Both midlist and backlist have largely vanished in recent times—partly because of the "Thor Power Tool" court decision, which turned what used to be a tax advantage into a tax liability; but partly, I and many other people think,* because of a widespread shift in publishers' priorities.

  In recent times, thanks largely to certain movies and television series, publishers have discovered that some things labelled "science fiction" can make lots of money. They have put more and more of their efforts and resources into those, at the expense of everything else. Thus we have "science fiction" shelves largely dominated by media spinoffs, "shared worlds," and an endless stream of trilogies. Many of these are written by highly capable writers who try hard to make them worthwhile—but they are not what these writers would *prefer*

---

*See, for example, Robert Silverberg's essay, "Gresham's Law and Science Fiction," in the opening pages of the anthology *Nebula Awards 31*.

to be writing, and the time spent writing them is not available to work on the much more original things of which they are capable.

We have editors and publishers who say things like, "We buy authors, not stories," and apparently see nothing wrong with that. Few book editors have enough autonomy to buy books on their own best judgment; they must first sell each one to a marketing committee who are unlikely even to read it.

We do *not* have the diversity that we once did and many of us would still prefer. Writers whose work appeals strongly to a well-defined but relatively small market segment no longer have an outlet; the readers who constitute that market segment have no place to find what they really want. Writers get a few chances to produce a best-seller; if they don't, they then sell nothing or they write media spinoffs to pay the rent. Readers learn to love the "lowest common denominator" books or they do without; those with minority tastes are not worth publishers' attention. Which may, in the long run, prove a very shortsighted attitude. As Silverberg says, ". . . modern-day publishing's emphasis on the bottom line seems to be killing science fiction as an adult genre. I loved *Planet Stories*, sure, but I doubt that I would have stuck with SF past the age of fifteen or so if I hadn't been able to move on to John Campbell's *Astounding* and Horace Gold's *Galaxy*."

- And, of course, we do have clear examples of accountant mentality in government, such as the recent movement to sell off national parks as a source of revenue.

Any individual, business, or organization, up to and including nations, must, of course, be fiscally prudent. But it is a serious mistake to make the bottom line the *only*, or even the primary, consideration in deciding how to live or do business. Maximizing the bottom line is fine, provided it's done in a way consistent with other goals such as living the way you want to and doing things you can be proud of. But that will not always maximize the bottom line, and maximizing is not always necessary.

Sometimes it's just necessary to have *enough*, not everything you conceivably could have. And if you have a good bottom line, you can afford to do some things that don't add a lot to it, simply because they contribute something worthwhile, or even admirable, to humanity.

Such sentiments are hardly new or original, of course, though they have been largely forgotten. For a particularly eloquent exposition that's been around for quite a while, I commend to you Eric Frank Russell's classic story "Late Night Final," which first appeared in *Astounding* in December 1948. It won't be easy to find, of course; in today's publishing climate, it's unlikely that anyone will have it in print when you look for it. But it's worth seeking out. And though the story was about a military invasion, the character Meredith's observation applies just as well to those who devote their lives to economic conquest: "If I, personally, were in complete possession of all the visible stars and their multitude of planets I would still be subject to one fundamental limitation, in this respect—that no man can eat more than his belly can hold."

Russell's story was about a civilization that had learned that lesson and thereby gone beyond the stages of succession I've talked about here. Can ours?

Working to Live,

or

Living to Work?

Bag Limit

My local newspaper recently ran a feature article headlined, "The Great American Bag Race," which I found both interesting and amusing in ways that neither the author nor the editor probably intended. The subject was the relative merits of paper and plastic grocery bags; the discussion included the reasons why many customers and grocers vehemently prefer one or the other, and the fierce economic competition between manufacturers of both.

Just a few years ago, practically all grocery stores in this country routinely stuffed a customer's groceries into paper bags. In the early eighties, plastic bags began to replace them in some places. By the time I sat down to write this, the two competitors were running neck and neck, with roughly equal numbers of paper and plastic bags in use. Some stores offer only one or the other, but many have found it worthwhile to offer a choice. Some customers prefer paper because they stand upright and are biodegradable. Others prefer plastic because of their handles and waterproof character, and point out that while they don't biodegrade as well as paper, landfills tend to slow down decomposition anyway and plastic bags take up less room in them. Grocers tend to favor plastic because it's cheaper, and bags are the largest single supply cost for supermarkets.

The article I mentioned reached no clear conclusion about which kind of bag was better overall, but it made clear that *both* kinds of bags contribute to the ubiquitous problems of resource consumption and solid waste disposal. The difference between them in terms of environmental impact is one of degree—and, when you come right down to it, pretty trivial. Ironically, neither the author nor anyone quoted in the article even hinted that there might be another option that offers much more significant advantages over *either* kind of bag.

Could it be, I wondered, that despite all the currently fashionable talk about Being Kind to the environment, most Americans have little interest in finding answers that make a *real* difference? Or is it just that they've forgotten, and forgotten how to imagine, that such answers may exist, and may even be almost laughably simple?

Grocery bags, it seemed to me, are one of the more obvious examples of such a case. If you want to reduce the problems of depleting materials to make them and later finding ways to get rid of them, the choice is not between paper and plastic.

The choice is between *disposable* and *reusable*.

Almost all Americans have come to take it for granted that every time they go to a store, the clerk will give them as many bags (made of *something*) as it takes to hold whatever they buy. According to the article I've mentioned, for groceries alone this averages 143 bags per person per year. But I know a few Americans who, during times of depression or wartime shortages, made or bought durable *cloth* bags which they could take with them every time they went shopping—and kept using them for years, or even decades. Until recently (when American ideas of throwaway convenience began to seep across the Atlantic), generations of Europeans took it for granted that they must provide their own containers for their purchases.

So there's plenty of historical proof that it can be done—and the potential environmental impact is *enormous* compared to the rather laughable "choice" between disposable plastic and disposable paper. Every bag of *any* kind consumes materials and energy for its manufacture, and eventually uses some combination of space and energy to dispose of it when it's no longer considered useful as a bag. A family of four, using disposable bags at the rate cited in that newspaper article, will incur these costs and problems for some 5700 bags over the course of ten years. The same family could carry the same groceries in five or six well-made canvas bags that could easily last the entire decade.

Look again at those numbers: five or six bags, total, versus five or six *thousand* to do the same job. Choosing plastic or paper bags on every trip to the store might reduce resource consumption and disposal requirements by a few percent. Choosing durable bags that seldom have to be replaced can reduce them by three orders of magnitude.

Which looks like a real choice to you?

An immediate reaction to the radical suggestion that both kinds of disposable bags should be generally replaced by very durable reusables might be, "It would put bag boys out of work!" But of course it wouldn't, really. Bag boys can fill a bag that I bring in every week just as easily as they can fill a new one every week. (Actually, bag boys seem to be disappearing anyway. Relatively few stores in my neighborhood still have separate people to stuff bags—in part, at least, because automatic price scanners now allow one person to ring up items and bag them in one motion.)

A more serious objection might be all the less obvious people whose jobs would disappear. If you drastically reduce the market for bags, you will indeed reduce the need for people to make bags, and to collect and process raw materials for them. If you take measures to reduce the amount of petroleum you refine and the number of trees you cut, you will indeed reduce the amount of employment available in doing those things.

But is that a big *problem*—or a big *opportunity*?

When a person, or a people, gets very much accustomed to doing things in a particular way, it becomes easy to lose sight of what is a means and what is an end. Basic human needs, for example, include food, clothing, and shelter. Our society has, over a very long period, evolved a mechanism for satisfying those needs that depends on a complicated system of exchange. When you strip away the obscuring details, the essence of it is that nobody can have those things unless somebody does the work of producing and distributing them—so we employ people to do the work, and in return give them currency with which they can purchase the products they need. Employment is a means to two ends: for the individual, it's a way to keep food on the table and a roof over the head; for society, it's a way to insure that enough food and roofs are produced.

The problem—the *real* problem—is that most of us have come, at least subconsciously, to regard employment not as a means to those ends, but as an end in itself. Therefore when technology reduces the amount of work required to satisfy human needs, we react in a truly bizarre fashion: instead of inventing new ways to distribute needed things to people who no longer have to work as hard to produce them, we invent new ways to make work for people to do! Wasteful, ultimately self-destructive ways, like making and throwing away six thousand shoddy bags to do the work of six good ones—and in the process making millions of people spend most of their waking lives doing tedious things that don't need to be done at all.

Bags are just one example. The philosophy of "dynamic" or "planned" obsolescence has pervaded much of our culture, with products of almost every description being deliberately designed to fall apart in a relatively (and unnecessarily) short time. What if the "canvas bag" approach became a basic tenet of our general philosophy? What if it became common practice to build *everything* to last? What if the entire country suffered an epidemic of pride in workmanship, so that everything we bought had a useful life two or three or a thousand times what we now expect?

Well, you may say, that would be terrible. We'd have a perfectly appalling welfare problem. But is that *really* what we'd have? Imagine a complex of technological advances and/or attitude changes that reduce the total amount of work that needs to be done by 50 percent. That *may* mean that 50 percent of the workforce is thrown out of jobs and onto welfare—but only if you insist on being locked into a system that sees *N* hours of employment per week as something that Everyone Must Have. If you and your culture are imaginative and flexible enough, it may instead mean that everybody can stay employed, and still be paid enough to live comfortably, but only needs to spend half as much time on the job—and has that much more time free to do things he or she *wants* to do.

Yes, it requires some restructuring of society—but isn't it worth it, especially when you consider the alternative? And the question is not purely hypothetical. The numbers are, but the essence of the situation is very much the one we're actually in. The culture has even shown *some* signs of responding to it—the "normal" work week *has* been reduced significantly in the last century—but the response is painfully slow and awkward. We're still so committed to "full employment," even if the definition of the term has changed slightly, that we waste enormous amounts of energy and materials and human potential doing work that not only needn't be done, but *shouldn't* be done.

This can't go on indefinitely. We will *have* to find ways to cut the waste, and better ways to match goods to consumers, more realistically geared to real needs and present capabilities. The more quickly and smoothly we can achieve that, the better it will be for everybody.

It won't be simple or easy; but then, important, worthwhile things seldom are. One complication that will need to be considered is the fact that not all kinds of work can be redistributed—"spread thinner"—with equal ease. If we cultivate the habit of building things to last (and, as consumers, taking care of them so they last), this clearly suggests a shorter work week as the norm in manufacturing occupations. But what about teachers, doctors, research scientists, or artists? Such occupations require such special personal qualifications that there may not be many more people to share in them, so it's a little hard to see how their workload might be reduced. In fact, it might well tend to increase, with a growing population of people having more free time and wanting to learn new things to do with it. On the other hand, people in such occupations tend to *like* their work, and might want to continue devoting large amounts of time to it anyway. Might such conditions lead at last to more respect and better pay for good teachers? Might more people having time to spend on education even raise the general level of intellectual interests, abilities, and activities? Maybe, just maybe—if things are handled right.

What about progress? If most things are built for durability, might this make it harder to sell new, improved devices, and thereby destroy much of the incentive to develop them? At least some of that effect is probably inevitable, but I'm not sure how much. Nor am I sure it's entirely a bad thing. Much as I admire and appreciate worthwhile innovation, it sometimes seems that much of what actually happens is a frantic and rather mindless chasing of fads, a perennial pursuit of The Latest Thing for no better reason than that it *is* The Latest Thing. The result is that most Latest Things come and go quickly and make no lasting contribution except to landfills.

Maybe it wouldn't be such a bad thing to slow change down just a bit, to take at least a temporary breather and give people a bit more time to learn to live *well* with what they have rather than constantly hurrying on to something newer that they dimly hope might be better. Besides, many of the real advances happening now deal with information handling and are not very materials-intensive. I see no intrinsic reason why we couldn't move toward sturdier houses and cars and clothes—and grocery bags—and at the same time toward better and better information systems.

In any case, it's high time more people were looking beyond short-term difficulties and trivial choices to see how we might best take advantage of the future. Often—and certainly in many matters that face us now—a situation looks like a "problem" rather than an "opportunity" only because we're not imaginative enough about "solutions."

Haste Makes Haste

It's a familiar concept in electronics—so familiar that everybody, not just engineers, has experienced it. It would be hard to find a person who hasn't heard a public address system produce nerve-wracking squeals or howls when the person at the microphone began to speak. Most of us even know the phenomenon by the name "feedback," though we may not know exactly why it's called that or how it works.

It's actually pretty simple. Feedback simply means that the output of some system, such as an audio amplifier, is fed back to the input so that it becomes part of the input. To take a very simple example, suppose you're feeding a simple 400-hertz sine-wave into an amplifier that produces an output with the same waveform as its input, but ten times the amplitude. This is useful, for example, if you're trying to let everyone in a large room hear the output of a low-power signal generator.

Now suppose you take part of the output signal—say, a tenth—and feed it back so that you add it to the input signal (which is still a 400-hertz sine-wave). Your actual input is now twice what it started out as, and if the amplifier multiplies that by ten, your output becomes twenty times the original input. Feed back a tenth of that and add it to the already doubled input, and you now have a quadrupled input being multiplied by ten . . .

Each cycle of feedback increases the output more and more, and it doesn't take many cycles before the loudspeakers are howling at the maximum level the system can produce. In the familiar practical situation, of course, the input is not a simple sine-wave; and the feedback happens not through a direct electrical connection, but because a microphone is too close to a speaker. But the waveforms of speech stay close enough to the same long enough for several cycles of feedback to occur, quickly building up the familiar and highly unpleasant howl.

Feedback is not necessarily bad; engineers find many useful applications for it. In my own thesis research, using the Mössbauer effect, I had to repeatedly cycle a moving sample through a range of very small, very precisely known velocities. The key to doing so was negative feedback: a sensor generated a signal proportional to the actual velocity, which was fed back and electronically *subtracted* from a signal proportional to what the velocity was supposed to be. The driving force was determined by the difference between what was and what should have been. If they were equal, no force was applied; if there was a little discrepancy, a little force was applied; if there was a big discrepancy, a big force was applied.

The problem in the Howling PA Effect is not that feedback happens per se, but that the feedback is *positive*. The amplifier is supposed to multiply the input signal by a certain amount, but if you keep feeding part of the amplified signal back to augment the input, the output quickly grows to excessive and undesirable levels. To put it bluntly, things get out of control, and what started out as something useful becomes a torture.

Like many physical principles and effects, this one has cultural analogs. A catastrophe like the Holocaust can happen because a charismatic demagogue feeds destructive ideas to a population that accepts and reiterates them, amplifying them from the beliefs of one person to the beliefs of thousands. The support of those thousands reinforces the original signal, encouraging the demagogue to spread his message still more aggressively, winning still more converts, who give him still more support . . .

Some tyrants are more subtle, but nonetheless encouraged by positive feedback—like the one in my title. No, it's not a typo. We've all heard "haste makes waste," but it seems to be just as true that haste makes still more haste, which makes still more haste. . . .

And pretty soon things are out of control and haste has become a goal in itself, driving people into an ever-increasing spiral of trying to do everything as fast as possible, whether it needs to be done that fast or not. The result is what I call "the ironic epidemic": a world full of people with more labor-saving devices than anyone before them, working longer hours than their parents, constantly complaining of stress, and forgetting how to relax. The latest manifestation to come to my attention was a Gannett newspaper with the subhead "Technology keeps more Americans wired to the office—even vacationers." The gist of it was that beepers, cell phones, faxes, modems, and other such devices have made it so easy for people to keep connected to their offices that they seldom, if ever, disconnect.

At first glance the problem appears to be, as the subhead on that article suggests, one of technology getting out of hand. I've actually heard people describe it in such terms as "technology has let us down," as if technology were some heavy-handed tyrant that forced us to use it (and be used by it) in certain ways. But that view misplaces the blame, and hides the possibility of a remedy.

The truth is that we do it to ourselves—and we, and only we, can stop it.

Oh, it's not each of us personally, though it is partly that. Not everyone who feels compelled to call the office or spend an hour online doing business every day of "vacation" does so because of a deep conviction that that's the best way to live. Many do it because their employers demand it and they feel they can't afford to refuse. Others do it for fear that if they don't, somebody else will, and thereby zoom past them on the corporate ladder. There's some truth, regrettably, in these views; some employers do feel perfectly justified in demanding far more work than they're paying for, and some employees will do whatever it takes to outcompete their colleagues.

And then there's the fact that some people become so hooked on new technologies that they insist on using their full capabilities whether they're needed or not—and, worse, demanding that those they deal with do so, too. E-mail (when it's working) is so easy that the volume of it now sent far exceeds the volume of correspondence when sending a letter required folding a piece of paper, stuffing it into an envelope, stamping it, and taking it to a mailbox. The speed and convenience of e-mail is a real blessing when complex matters need to be discussed and resolved quickly.

But since it can be sent and answered very quickly, too many people now use it to send messages that aren't really worth sending, and expect everyone to answer immediately, whether there's any actual reason for haste or not. (One result, ironically, can be extra long delays in response. A while back I went through a spell of having a lot of trouble getting outgoing e-mail messages to go anywhere, and people took it so much for granted that I would reply instantly by e-mail that they didn't supply any other kind of address. So before changing providers, I acquired quite a backlog of messages that I couldn't answer because I had only an e-mail address and an unreliable e-mail system. Moral: It's a really good idea to include a snail-mail address even when using e-mail!)

That kind of delay, of course, is a bit unusual. The more usual result of the recently exploded prevalence of e-mail is that many of us are receiving unprecedented quantities of written communication and feeling pressured to answer it all Right This Minute. Which contributes to lots of other people having even more of the same problem . . .

Positive feedback does it again.

Is there a way out—a way to get the advantages of ever faster and easier communication without becoming enslaved by it? Of course there is, but the responsibility for making it happen rests squarely with human beings. The gadgets are just tools; it's up to us to decide when and how to use them, and what demands we are willing to let other human beings make on us because they're available. The technology, in and of itself, does not force us to use it for everything it can do. We need to learn to ask ourselves, "Is this job really worth doing?" and "Is this tool the best one for the job?" and "Does this really have to be done right now, or can it wait?"

Employer pressures are not quite so simple, of course. If employers are making unreasonable demands and most employees go meekly along with them, a lone individual who sees the insanity of the accelerating spiral may have to go along or lose his job. But if lots of employees refuse to go along, it gets much harder for employers to sustain their demands. Such considerations are why unions were invented—and employers who shudder at the excesses of which unions have sometimes been guilty would do well to remember that.

In time, I suspect, reaction to current excesses, stresses, and strains will lead to some kind of moderation and a new set of Lines Not To Be Crossed. Vacations—real vacations, not "on-call" vacations—serve a battery-charging function that benefits the only thing some employers understand: the Bottom Line. Employees simply work better if they periodically get completely away from the work. And the Bottom Line is not the only thing that matters. Ultimately people don't live to work; they work to live. And you won't have to look far to find a psychologist to tell you that life has to have more in it than work. But when technology makes continuous, intense work so feasible, people—both employers and employed—are going to have to learn when and how to say no.

Actually, this problem has occurred and been dealt with before, and I don't mean just recently. There's a very old concept in some religions called the Sabbath: one day every week when virtually all work halts and people are free to concentrate, singly or collectively, on other aspects of life. In our present culture at large, this concept has largely been lost. These days Sundays and holidays are seen by many as times for big commercial promotions. There is essentially no time when practically everyone is free to go off alone, or gather with friends and relatives, to do "non-work" things.

This is not entirely a bad thing. When everybody has the same day off, many places get unpleasantly crowded on that day. And some religions' Sabbath observances were less matters of freedom to recreate than rigidly enforced ritual. But there was an important, beneficial idea at the heart of the Sabbath, whether religiously or secularly motivated, that has largely been lost. It may be an idea whose time has come—again.

The key to getting it back, I think, will be the growing, deep-seated understanding that we do not always have to do something just because a technology makes it possible. What we need to learn was very well put by *Analog* reader Gwen Ross: "Intelligence sees how to. Wisdom sees when to."

Our Environment
and Us

Megachoice

One of the most controversial things *Analog* has published was not a story, not an editorial, but two paragraphs one of our authors wrote in reply to a "Brass Tacks" letter commenting on a science fact article.

Paul Birch's article, "A Visit to Suprajupiter" (December 1992), described a grandiose and imaginative scheme for building an artificial habitat completely surrounding Jupiter that could be inhabited by vast numbers of people. Reader John Vester, in his letter published in August 1993, applauded Suprajupiter as an imaginative exercise but expressed some concerns about *why* such a project might be undertaken, a question not given much attention in the original article.

As we often do in such cases, we forwarded a copy of Mr. Vester's letter to Mr. Birch in case he wished to run a reply when we printed it in "Brass Tacks." His seven-paragraph response included these comments: "I believe we must fill the Universe and subdue it, because God made it for us and commands us to." And, "As for finding a way out of the Malthusian trap, we were never in it. . . . Even on Earth, current technology would allow population to grow by *at least* a further factor of 40,000. . . . The terrestrial environment is nowhere near destruction; we've *not* spoilt it, we've improved it. The so-called limits to growth are nonexistent."

And then the fun began.

Those two paragraphs drew more mail than many whole stories or articles. Some readers were shocked that we would print such statements. One believed that they destroyed the credibility of anything else the author ever said. Others disputed them heatedly and in detail. Some wondered whether they were even meant to be taken seriously, or whether they were said with tongue in cheek to provoke thought and discussion.

Since just such provocation is well known to be one of *Analog*'s most characteristic (and useful) activities, I would never have considered suppressing Mr. Birch's comments, no matter how much or how little I agreed with them. Nor do I believe, even if they are utterly wrong, that they destroy every shred of his credibility. Everything *anybody* says is subject to doubt and question. Nothing should be considered automatically believable because a Good Source says it, and the utterance of one piece of sheer nonsense does not invalidate every sound statement a person has made (though it may *reduce* his credibility). Most of us have uttered nonsense more than once, whether intentionally or un, and most of us would be in sad shape if that meant nobody should ever listen to us again.

I frankly don't know whether Mr. Birch seriously believed everything he said, nor do I care. I asked for his response to a letter commenting on his article; I printed it; and I'm glad to see it stirring up heated discussion.

But I do think his points are open to serious debate and criticism—and, sadly, I don't think that, by their very nature, that debate can be resolved to everybody's satisfaction. Because a crucial part of the answer involves personal preferences—tastes—and a course of action that pleases people with one sort of tastes will be violently displeasing to those with another sort.

Point 1: Mr. Birch, like anyone else, may believe whatever he chooses about what, and for what reasons, "God made . . . and commanded." He *cannot* expect such arguments to carry any weight at all with people who don't happen to share those beliefs. So while they may explain his personal advocacy of a particular course of action, they in no way justify anyone else's following that course. It may be a good one, but if so, it needs other justifications.

I found myself thinking of this part of the argument during a Westercon panel on morals and ethics in science fiction, when a member of the audience raised an interesting and pertinent question. Even if you do accept the "God commanded . . ." argument, he asked, "If God said, 'Be fruitful and multiply,' did he mean *forever*?"

Point 2: Hardly anybody these days believes that there are *no* limits to growth, or that Malthus's basic idea had no merit at all. He was wrong about the details, but even the most optimistic proponents of highly advanced technologies like nanotechnology—or at least the ones who've taken a close look at numbers—concede that the basic idea was right: the population of any given system *cannot* grow indefinitely. The limits are farther out than Malthus thought, but that doesn't mean there are *no* limits.

Most of the controversy Paul Birch has provoked centers on his specific claims about the system all our current (known) readers inhabit: the one we call Earth. Can it really support more than 200 trillion people? Even if it can, *should* it? Has human activity really improved our environment more than it has messed it up?

On the last question, we run smack up against those questions of taste. Some people *like* living in densely populated cities; some hate it. On the first question, we're dealing with a conjecture on which people of many persuasions would like to believe more than they can prove.

One comment that came up on that morality and ethics panel expressed a common belief that may contain too much wishful thinking to be a solid argument. It maintains that we *can't* run this planet's population up that high because we're too dependent on all the other species that live here. Wipe out too many more of them, the argument goes, and *Homo sapiens* will soon follow.

That may be true—or it may not. Personally, I would like to preserve a high level of diversity, but I'm not at all sure that that particular argument is strong enough to convince people who disagree with me. Very probably, the spotted owl, in and of itself, is *not* necessary to our survival. Neither is the snail darter. Our present ecosystem is not the only possible or viable ecosystem—or even the only one containing our species as an important component. As a hard science fiction writer, I find it uncomfortably easy to imagine a future in which our descendants *have* wiped out almost all other natural lifeforms, and developed a new, simpler, entirely artificial ecosystem designed for the sole purpose of supporting vast numbers of *us*. I don't think anybody *now* understands ecology well enough to do it, but I can imagine enough being learned to make it possible in the future.

But I don't for a minute think that it *should* be done.

Why? With such questions, quite frankly, we're getting down to the level of ethics that depends on principles accepted without proof—the moral equivalent of geometrical axioms. It's as hard to convince somebody logically of those as it is for Mr. Birch to convince a non-believer that everybody should do what his God allegedly commanded.

So I'm reduced to saying that maybe we *can* do away with most other lifeforms, but we shouldn't anyway—if only because it would be a nasty thing to do. As the naturalist-philosopher Joseph Wood Krutch wrote in 1955, "Albert Schweitzer remarks somewhere that we owe kindness even to an insect when we can afford to show it, just because we ought to do something to make up for all the cruelties, necessary as well as unnecessary, which we have inflicted upon almost the whole of animate creation. Probably not one man in ten is capable of understanding such moral and aesthetic considerations, much less of permitting his conduct to be guided by them. . . ." That number is probably a bit higher now, but needs to be still higher if diversity is to be preserved.

The argument that we *need* all those other species to survive is simply too shaky to rely on, so those of us who believe we need them for other reasons need other arguments. A large number of people simply believing that diversity is intrinsically desirable, and that ethical considerations can apply not only to our treatment of other *human* beings but to our treatment of other *beings*, may be the best we can do.

In any case, returning to the original point, the answer to the question of whether Earth *can* support the human population Mr. Birch claims may well be *yes*. On whether it *should*, he and I pretty obviously differ. I've already discussed one reason; others bring us back to the realm of personal taste.

Mr. Birch says flatly, "We've not spoilt [the environment], we've improved it." A good many readers disagreed vehemently. What they're all really saying is: Mr. Birch likes it better than he used to; some other people like it less.

My own answer would be mixed. I certainly prefer my present life expectancy to the one I would have had in any century before this one, and I'm grateful that I now have the ability to get around and experience firsthand far more of the world's diversity than most of my ancestors could. On the other hand, I'm sorry to see so much of that diversity being homogenized and paved over, and to see so much of the population crowded into such tight spaces that they're living in squalor and clawing at each other like rats in too small a cage.

If you read over that last paragraph, you may notice an interesting and significant pattern in what I like and don't like. In general, the things I consider steps backward are directly associated with high population density. It's true that I am, by choice, perhaps influenced by where various parts of my upbringing were spent, primarily a country boy—but I am not unique. I realize that some people prefer living in densely populated areas; and even I freely admit that cities produce not only conditions I dislike, but things that I like very much and can find nowhere else. Great museums, libraries, theaters, orchestras, and restaurants are very much products of cities, for reasons that are easily understood in terms of our past history. (They may be less pertinent in some possible futures, but that's another essay!) But I would rather live in the country and make an occasional trip to the city to enjoy those things, than live full-time amid congestion and noise and pollution to be near them.

So would a great many other people. The observation that so many people who can afford to prefer to live outside urban congestion is, I think, very relevant to Birch's vision of an Earth of 200 trillion.

We might well be able to do it, and some people would love it. Others would not—but they would have no choice.

Tastes differ: some people like high population densities; some detest them. This is not the same as liking or not liking people. I tend to like people, as individuals, when I get to know them; I do *not* like crowds. I like people; I also like jalapeño peppers—but I don't think the best thing that could be done for the world is to fill it with either.

My biggest problem with Mr. Birch's vision of that superpopulated Earth, I think, is that it leaves very little room for diversity—not just of species, but of human tastes and lifestyles. I refer specifically to this matter of preference for high or low population density. Our present Earth, we know, contains people of both extreme types, and everything in between. It offers them all at least a modicum of choice. If you're stuck in a city and long for quiet and elbow room, there are still reasonably open spaces to which you can relocate—maybe not now, but you can work toward it for the future. If you grow up bored stiff on a farm and crave company and excitement, you can move to the city.

Mr. Birch's vision, as I understand it, would eliminate that choice. If you populate a planet to its limit, whether that limit be 200 trillion or some other figure, pretty much everybody is going to be living under the most crowded possible conditions. And to me, that is a very, very serious flaw in such a vision as a goal for the future. My world has room in it for people who prefer his kind of living conditions; his has no room for my kind.

You choose.

The Forever Fallacy

When I was growing up, I sometimes heard an adult on the radio or on a stage speak solemnly of "the everlasting hills." If you were like me, this probably struck you as carrying poetic license a bit far. Even as a child, I spent enough time in encyclopedias and natural history museums to know that hills are not everlasting. They just change a lot more slowly than people, so we don't usually see the changes. But we do see the evidence that the changes happen, and are often quite drastic. *Nothing* is everlasting—not even the planet on which those hills are mere bumps, or the solar system or galaxy of which it is a part.

Understanding that, I was even less able to take seriously claims that anything human-made was "everlasting." When I learned how much of the world's wilderness had been sacrificed to make room for civilization, and how deeply satisfying time spent in the few remnants could be, I *wanted* to believe the reassurances I constantly heard from teachers and politicians. National parks and places like the Adirondack Forest Preserve in New York State, I was told, guaranteed that these scattered reminders of the past would stay "forever wild." I would have loved to believe that—but deep down, I knew that "forever" in this case meant even less. "Forever" is something that many humans seem compelled to seek; but when you think you have it in hand, it is, at best, a comforting illusion.

The "everlasting hills," at least, usually last many human lifetimes. But "forever wild," I've known all along, means "until the politicians in power change their minds." That *might* be a lifetime or more—or it might be tomorrow.

Or, as it now appears, *today*.

Washington is currently infested with politicians who think a good way to move the budget closer to balance (and win points with their most useful constituents) is to reduce or eliminate funding for things like national parks, or even to close them down or sell them off to private interests for more "profitable" use. We have people in positions of power over national parks who can say of them, apparently with straight faces, things like, "If you've been there once, you don't need to go again."

Well, maybe *he* doesn't, but I do. So do a lot of other people. And it just may be that he needs it at least as much as any of us, whether he's capable of understanding that or not.

It's true that parks and wilderness preserves are not great producers of income measured in dollars and cents. That's not their job—but their job is at least as important as the generation of Wealth. Their main values are therapeutic and educational. I can testify from many, many personal experiences that periodic immersion in wilderness can be one of the best treatments to restore peace of mind sorely tried by the more exasperating aspects of civilization. And it can serve as a sobering and salubrious reminder that this planet does not exist solely for our convenience, and that everything on it—including us—owes its life to a complicated set of interactions with other living things.

To someone who has spent enough time in wilderness to develop some understanding of it, or even one who has grown up on a farm, that may seem too obvious to mention. But it isn't. For a city kid who has never been to the country before, the awareness of being part of something much larger may be not a reminder, but an astonishing revelation of something previously unsuspected.

Valuable as it is for many things, technology makes it all too easy to forget our relationships to other things. Plenty of city dwellers have little or no grasp of the fact that meat *really* comes from animals and vegetables from "dirt." At least at the gut level, they may associate food with nothing more distant from themselves than a plastic package on a store shelf. Anyone with no more understanding than that is living a life based on a dangerous delusion. If such a person votes or holds office, his or her delusion endangers *all* of us.

You may find it hard to believe that such an extreme form of ignorance exists, but I assure you it does. Subtler variations of it are far more widespread than you might think. I've known people who were brilliant practitioners of one area of science and technology, yet had little or no grasp of ecology—a dangerous situation for people engaged in activities likely to have far-reaching ecological consequences. It may not be their fault; they may simply not have had enough exposure to the ecosystem beyond their own walls to make them think much about it.

Science fiction people place a high value on "sense of wonder," and sometimes bemoan the lack of it in recent examples of their favorite literature. But these same people sometimes get so wrapped up in *technology*—the joy of doing nifty tricks with fancy gadgets—that they lose sight of the sense of wonder in *science*. Technology is an outgrowth of science that enables humans to do things that they couldn't do without it, some of which are indeed quite wondrous. Science is the study of *all* the possibilities in the universe, of which technology is only one subset.

It's possible to get so enthralled by the wonders of technology that you lose sight of the wonders that don't depend on it. Our planet, for example, is an enormously complex and astonishing thing, and it was so even before people started making big changes in it. A large part of the fun of science fiction is imagining how many kinds of *other* worlds might exist, as big and multifaceted as this one—or even more so. Most of us now live in such small, artificial microcosms that we've forgotten—or never learned—just how big, diverse, and remarkable our own world was even before we got our hands on it. The Kyyra, in my novels *The Sins of the Fathers* and *Lifeboat Earth*, had turned whole planets into such artificial microcosms, and were quite overwhelmed by the spectacle of a relatively unaltered planet.

For those who *do* appreciate the magnitude of what natural forces alone have accomplished in creating a world like Earth, the realization may or may not take a religious form. I remember somebody asking the manager of a remote wilderness camp in Africa whether the isolation ever got to him. "Do you miss being able to go to the store," they asked, "or to church?"

"I'm *in* church," he said, gesturing inclusively at the Serengeti all around us, with its distant horizons and incredible abundance and diversity of life.

And then there was my good friend who looked out from a high ledge in the Rocky Mountains and remarked, "I don't know if there's a God or not—but if there is, He's a hell of a guy!"

Whether or not you see it in theological terms is, it seems to me, a matter of taste, and ultimately not very important. What *is* important, and profoundly so, is that we all need to have some understanding of the fact that there's a lot more to the world than us, and all the pieces are related in a rich tapestry of ways.

Parks and wilderness preserves provide opportunities to see that tapestry close up. Some might object that not everybody can use them directly, but that hardly justifies doing away with them. First, even those who can't get there can get *some* benefit just by knowing that they're there, by reading or watching television shows about them—though that is, at best, a pale substitute for the real thing. Second, if actual visits to the park have potential value to many people, and not all of them are able to take advantage of it, that doesn't mean we should take them away from those who are already using them. It means we should try to find ways to get those benefits to more people. And since the experience, by its very nature, is ruined by crowding, that means we need *more* parks and wild lands, not fewer.

All these arguments, of course, are based on the value of wild lands to *us*. You might object that that's yet another example of the arrogant anthropocentrism that has already gotten us into some trouble and holds the potential for much more. Privately, I would agree, at least partially—but sometimes you have to work with the materials you can get.

I tend to agree with naturalist Joseph Wood Krutch's observation (quoted in the previous essay) that we have purely moral obligations to other living things, even though few people are capable of understanding such concepts, much less acting on them. "But," he added, "perhaps twice as many, though still far from a majority, are beginning to realize that the reckless laying waste of the earth has practical consequences. . . ."

And therein lies a hint of hope. Certainly the "everything for a buck" crowd in Washington has flunked the "Krutch test." I don't think it likely that anybody will change their thinking enough to pass it, but it just *might* be possible to convince some of them that "what's in it for us" can include a few things not measurable in dollars and cents—though even that will be an uphill battle.

The value of parks and wild lands is *not* primarily financial. There are those, some of them in positions of high power, of such short and narrow vision that they sincerely believe that the value of all things can be measured by the current quarter's bottom line. They, perhaps, need the parks' real benefits most of all, though many of them may be beyond hope of realizing it. The idea that parks and wild lands are always more valuable as sources of salable timber and minerals misses the whole point. But if their fate is in the hands of people who can't grasp that point, we'll have to justify their preservation in terms those people *can* understand.

Space enthusiasts sometimes point out that moving much industry, including mining, into space can and should reduce the pressure on open land and make it possible to preserve more of it. This may be, eventually, but unfortunately we're not very close to having that situation *now*. While we're waiting, and some are fighting another uphill battle to achieve *that* state of affairs, the pressure to mine and log and build on every possible acre goes on. Space may *eventually* reduce the need for Earth-based mining, and more sensible attitudes toward population and economics may eventually reduce the tendency to build more and more housing and shopping centers. But a torn-up wilderness cannot be quickly fixed. "Wrecked" isn't forever, either, but it's close enough on a human scale. Drained wetlands and clearcut forests can take decades or centuries to fully recover.

Fortunately, elective office isn't forever, either. . . .

If they're doing what they do best, national parks and protected wilderness *won't* turn a profit—unless, perhaps, you charge enormous admission fees. But that turns them into another exclusive toy for the rich; and while many of the rich desperately need to learn the things they can learn from wild lands, they are by no means the *only* ones who need it.

All of us need to be reminded periodically of where we came from and how we fit into a larger scheme—including, most emphatically, those who make laws and public policy.

Public vs. Private

In "The Forever Fallacy," I commented on the trap of believing that anything lasts forever—especially anything that depends on politicians keeping their promises. My main example was a movement afoot in the federal government to sell off portions of national parks and other public lands to private enterprise. Many readers largely agreed with my comments, but a number of them (respectably if not impressively into the single digits) attacked me in no uncertain terms for "defending the socialized park system."

When I wrote "The Forever Fallacy," it never occurred to me that I was doing anything so grandiose. My intent was merely to point out the threat I saw in a move that seemed likely to destroy something that needs to be preserved. But those readers' comments did raise an interesting question, particularly since in large measure I agree with them in principle. It's quite true, as one of them pointed out, that the concept of individual liberty coupled with personal responsibility has long been (and will continue to be) an ideal that has shaped much of the thinking in *Analog*'s pages. So how can I object to privatizing *anything*?

The difference between my critics and myself, I think, is that their thinking is dominated by one principle, with little or no regard for others. I feel strongly about several principles, of which individual freedom is high on the list. But I am not so singlemindedly fixated on any one of them that I'm unwilling to consider others at the same time.

Personally, I'm an anarchist at heart—but that does not mean I'm a "terrorist" or advocate violent overthrow of the government. I simply agree in principle with whoever it was (Jefferson?) that said, "That government is best that governs least." My idea of Utopia is a place where everyone is so wise, benevolent, and responsible that no government is necessary to insure smooth functioning of the infrastructure and ample opportunity for all to create safe, comfortable lives in accordance with their own tastes and principles.

Unfortunately, I don't live there. In the real world as we know it, I fear, a certain amount of government is a textbook example of "necessary evil"—with equal emphasis on both words. Furthermore, while private enterprise and commercial competition do some very important things very well, they do not necessarily do *everything* better than any other possible method. And sometimes getting things *done* is more important than *how* they get done.*

To answer the critics who fear I've turned from a (lower-case) libertarian to a (lowercase) socialist because I've criticized a proposal for the government to privatize one specific thing, I must respond on two levels. First, and simpler, is the pragmatic question of what seems the best or least harmful course of action *within the limits of the world as it exists or as I can reasonably hope to change it.*

Given the reality that I *will* be taxed, parks and wilderness preserves seem to me one of the more worthwhile ways my taxes can be spent. I think they serve important functions, not just for me, but for the country at large. Sometimes I think they're one of the few government programs that *are* worthwhile—and if I *must* pay taxes, I'd rather see them spent on something worthwhile than wasted. No, I *don't* like someone else deciding how to spend my money.** But given that it's going to happen anyway, I am less unhappy with expenditures that do something important. Parks are important, we now have them, and selling them to loggers seems an improbable way to preserve them.

---

*No, I'm not saying, "The end justifies the means"—though it often does, to a considerable extent. The reason that slogan has fallen into disfavor is that it doesn't make allowance for the fact that, "You can't do just one thing." That is, whatever you do has side effects in addition to the one you are after. Those are part of the real "end," and so must be taken into account in evaluating options.

**I'd love to see somebody try the system suggested in the story "We, the People," by Jack C. Haldeman II (*Analog*, Mid-September 1983), wherein people pay taxes but specify how much of them is to be spent on what!

The second, and more complicated, question is whether the system itself is a good one. That is, even if preserving wild lands is important, is government ownership and control of them a good, or even a defensible, way to do it?

One reader objects to it, and to "*any* socialist mechanism," because it demands support under threat of force from those who would rather not support it. This reader has a carefully thought-out system of ethics founded on the principle that force (in his word, "violence") for any reason except self-defense (or assisting someone else's defense of person or property) is wrong and must be avoided at essentially any cost. I can sympathize with that; it's strongly akin to my own fundamental moral principle, that people should be free to do whatever they want as long as it doesn't hurt anybody else.

However, I don't think that direct physical violence is the only way to hurt somebody. Nor do I think other *humans* are the *only* thing that should be considered in making ethical decisions. I may never see the neighbor who lives a mile downriver from me, but I can certainly hurt him by dumping barrel after barrel of toxic waste into it. He is quite justified in making me cease and desist, forcibly if necessary. And while either of us might reasonably catch as many fish as we and our families can eat, neither of us needs—or should expect to be allowed—to slaughter vast numbers of them indiscriminately, *whether or not* other humans will later need them for food.

The principle of "anybody can do anything except violence" works well as long as no individual has very much power to affect other people or the environment. Conceivably it *could* work well even then, if everybody could be trusted not to abuse their power—but history has abundantly demonstrated that, in the world as it has existed so far, they can't. A regrettable side effect of the increase in individual power brought by technology is that it becomes harder and harder to leave people entirely to their own judgments. The same technologies that enable individuals to do more desirable and beneficial things also enable them to do lots of damage with unprecedented ease, and it's Not Smart to let them do so. If preventing at least the more extreme possibilities requires some measure of public control, that may be the lesser evil.

However, please note: If no coercion is allowed, *no* public works are possible. If everything is private, you're completely dependent on private good will, competence, and coordination. In theory, at least, it's possible to have a road system entirely in private hands, financed by tolls. Of course, you may have to stop and fork over a new toll on every road you turn onto, and you may not be allowed to turn onto some of them if the owner doesn't like you. In principle, I suppose, that's the way it *should* be, if nobody's allowed to coerce anybody.

Of course, drivers may object that the system's unfair because even though they're the *direct* users, they're not the only ones deriving benefit from the roads. Anyone using anything trucked in from outside his immediate neighborhood is indirectly using them and should pay some of their costs. Right, say the road owners to the truckers—so build something into your prices to collect a fair share of the tolls from your customers.

There are no doubt some who think such a system would be wonderful. I think it would be such a confounded nuisance as to outweigh its theoretically superior fairness. I don't give up freedoms lightly; but after careful consideration, I am less bothered by paying a reasonable tax to support an extensive public road system than I would be by the constant aggravation of paying tolls to a zillion road owners and wondering which ones will refuse to let me pass even if I pay their price.

So: Yes, I am basically libertarian, but my experience strongly suggests that for a few things public or government control has enough advantages to make it worthwhile anyway. If you agree (and I realize that some never will), then you're stuck with allowing a limited amount of coercion. The question then becomes, when and why is such centralized control justifiable?

I've already said why I suspect roads are one example: The bookkeeping for a private substitute is likely to be more trouble than it's worth, and subject to individual abuses hardly distinguishable from those of corrupt government officials. Similar arguments could well apply to other transportation networks such as waterways and airspace. (Do you really want anybody to be able to try flying a plane around your neighborhood, with no assurance that he knows how?) Public education is another example, because everybody suffers if the country is run by the abysmally ignorant—and how many of the poor would get even minimal education if they had to buy it from private vendors?* Preservation of wild lands fits the category for related reasons: One of its most important functions is as another aspect of public education.

At least one reader who commented on "The Forever Fallacy" clearly doesn't care whether any wild lands are protected or the general population ever has a chance to see any. He says, "You want wild lands; buy some." He also mentions private organizations like The Nature Conservancy, which preserves land by buying it and maintaining it as wilderness. Good ideas, as far as they go; I've personally bought land for the express purpose of keeping it undeveloped, and I've been a member of The Nature Conservancy for years. But neither I nor relatively small organizations like The Nature Conservancy can do very much compared to the national park system. Yes, I *could* do more—if I wanted to devote all my time and funds to that at the expense of everything else I might also care about. But that's a highly unsatisfying solution, and a great many others would find it so, too.

Tough, says this reader; decide what's important to you and put your money where your mouth is. But he misses the point—several of them.

I didn't start all this just because I want a chunk of wilderness, but because I think a wilderness experience needs to be *available* for *everybody*—especially children. If at least some of them are exposed to something beyond the artificiality of cities, maybe they won't grow up as abysmally and dangerously ignorant of ecology as many present adults.

Only those who already know enough about wilderness to un-

---

*Yes, I know the public schools we have now leave much to be desired. That's a reason to improve them, not to do away with them.

derstand its value will be motivated to buy it. Only the relatively well-off will be able to. Those who do buy their own have only a little piece of one ecosystem, and may not choose to share it. Organizations like The Nature Conservancy can buy some, but generally not really large tracts—and true wilderness cannot be preserved in half-acre parcels.

Things like the national park system have two main values. They can preserve *large* expanses, big enough to house fully functional ecosystems. And they enable people who *don't* already understand wilderness and its value—e.g., casual vacationers—to discover it. They even let those "toe-dippers" experience not just the one little piece of an ecosystem that a non-wealthy person might own, but something of the *diversity* this planet has produced. A family that can save up time and money for one vacation per summer might see the Maine seacoast at Acadia one year, the Florida Everglades the next, the desert of Joshua Tree in another, and the geological storybook of Grand Canyon in still another.

Naturally, it's reasonable to expect the actual users to pay more than the nonusers. (I once complained emphatically when I found that Congress had inadvertently abolished user fees on a lot of federal lands that I was camping on, and was relieved when they were restored.) But until shown convincing reasons to do otherwise, I will persist in the belief that the public subsidy is worthwhile to keep the opportunity to visit places like Olympic and Denali open to any family that can save up enough to get there (subject, of course, to capacity).

Another reader seems to understand the value of preserving wilderness, but thinks private ownership will do a better job. But will it? To show how Private is better than Government, he suggests comparing a fast food restaurant (where you "get a product you want at a fair price") and a government office (where you "pay a high price for slow, rude service that you did not even want"). His comparison between those examples is valid, but the examples chosen have little to do with How to Protect Wilderness. A better comparison would be between a national park and Disneyland. The customers at both get what they want, but they're looking for different things. An even better, and very direct, comparison might be between Great Smoky Mountains National Park (which offers some of the finest backcountry camping in the country) and the garish honky-tonk of tourist traps that private enterprise has created just north of it.

One reader claims "corporations last longer than governments." A very few have, but the generalization is farfetched to say the least; huge numbers of corporations are born and soon die during the life of almost any country. He also says, "If you've been to a major western US park, you probably don't like what has already been done." Well, I've been to most of them; and while I'm not entirely pleased, I like them better as they are than if they'd been strip-mined or developed like Disneyland. And from all I've seen, those are the kinds of things most likely to happen to them if they were sold to private outfits. I'd like to believe otherwise, but until somebody gets a lot more concrete with suggestions about how private will do it better, I'm most reluctant to sacrifice what we already have. It's not perfect, but it appears to me the less objectionable of the *currently available* evils.

One of the more interesting comments I received was, "Presumably you also want more money for the nationalist socialist space program." This is presumption in every sense of the word; I said nothing about space, and I neither said nor implied that I prefer "nationalist socialist" approaches in general. However, since the subject has been raised, getting humanity into space seems to me another of those things so important to our species' future that they simply need to be *done*, and I don't particularly care who does it or how. Given my druthers, I'd favor private; and I'm pleased to see that a number of companies have finally realized there is money to be made in space and are working to do so. So far, though, despite its limitations, NASA has a much longer record of actual accomplishment. Until we're farther along, I don't mind seeing *both* approaches trying, and may the best system win.

Parks and wilderness preserves are similar in that I think we need them, *as parks and preserves*, and as long as we have them, I don't particularly care whether they're public or private. If somebody has concrete, believable proposals for how they can be privatized *without destroying their value as parks and preserves*, that's fine with me. But vague assurances that It Will Happen are not enough. Simple faith that Competition for Dollars Will Make Everything All Right is touching but unconvincing.

Competition for dollars does an excellent job of stimulating innovation in manufacturing—but it also stimulates many businessmen to cut corners in everything from product quality to customer service to treatment of employees because they care far more about the bottom line than about how they maximize it. The proposals I've seen for selling off federal lands provided no reason to believe they would continue to be managed as parks or preserves rather than golf courses or shopping malls, and I'm pretty sure most buyers would see those as quicker and surer ways to profits. Sure, privatize the parks, *if* you can guarantee that they will still *be* parks, run at least as well as they are now run by the government—but don't tell me that clear-cutting and strip-mining them is better just because it's done by Private Enterprise.

The "public vs. private" debate, as it's often presented, seems to me yet another of those "all or nothing" fallacies: the mistaken assumption that everything must be done *this* way or everything must be *done* that way, when in fact this way might work better for some things and that way for others. Sometimes getting the job *done* is more important than doing it according to somebody's pet ideology. *How well a knife cuts* is more important than whose name is on the blade.

Snakes or Paychecks: *Is* That the Question?

There is an area north of New York City—I hesitate to identify it too specifically, lest it attract too much attention from outside—that is highly prized by many people for its scenic beauty and natural diversity. This area, in fact, is considered by some to be the birthplace of the environmental movement in late twentieth-century America. Much that has happened since can be traced back to the protest that stopped plans for a power plant that would have meant seriously defacing one of the most scenic mountains in this area in the 1960s.

Well, they're at it again—both developers and protesters—and my observations of the process prompt me to comment on some of the factors involved in such cases generally.

These observations happened almost unwittingly; I hadn't intended to go to a political demonstration. I thought I was just going on a hike, one I'd seen announced in an area I was unfamiliar with but interested in. I usually avoid organized hikes, especially with large groups, but in this case it seemed likely to be worthwhile. Published information on trails in this area was unusually sketchy, so going with somebody who knew them seemed a good way to learn about them. Besides, I wasn't expecting a big group. Even though I knew the mountain in question was threatened with a large-scale open quarrying operation, my impression was that the group running the hike was a relatively small one in which I knew several members, and that the hike was just a chance for interested people to familiarize themselves with the area.

What I found when I got there was quite different. The group through which I'd heard about the event was only peripherally involved; a different group had organized it, its main purpose was to rally opposition to the mining permit, and there were many dozens of people there, most of them unfamiliar. There *were* hikes, all of which looked as if they were going to be too large for my tastes; but before that there was distribution of pamphlets and remarks by speakers from several environmental organizations, explaining what the proposal involved and why they hoped their listeners and fellow hikers would join them in opposing it.

And there were hecklers. Gathered at one side of the crowd was a group of mine workers wearing signs that said things like "SNAKES OR PAYCHECKS?" and "WHO WILL FEED OUR CHILDREN"? From time to time they interrupted the speakers with shouted jeers, and one of them tried to get his own time at the microphone. Those in charge declined, politely but firmly—and probably wisely. "This is our rally," one of them said. "You're welcome to hold your own at another time and place." Certainly the issue should be debated, but there would be other times and places set aside specifically for that. And when it happened, it would surely not be fast or pleasant. Letting it start during an hour of gathering before a hike was unlikely to be productive for either side.

But I was intrigued by the miners' signs and interjections, and I do think they warrant some comment now—though I doubt that my comments will be quite what they'd prefer.

First off, the choice is not as plain and clear-cut as "Snakes or paychecks?" Reducing things to such rudimentary and loaded dichotomies is simplistic sophistry and undermines respect for anyone who indulges in it—no matter *what* side they're on. Few of those fighting to preserve the mountain had preserving snakes as their only goal, or even an important goal—though probably most of them would cheerfully acknowledge that it would be a minor side effect of what they were really after. What they were after was saving not "snakes," *per se*, but the integrity of the mountain and the whole attached ecosystem—which happens to include snakes (among them one endangered species), but also includes a great many other things.

Second, the proper answer to "Who will feed our children?" should be, "*You* will." If you weren't prepared to do so, you shouldn't have had them—and nobody ever promised you that you'd always be able to do it the same way. Times are changing faster than ever, and most people can expect to have to change careers at least once during their lives.

In this particular instance, it isn't clear that anybody's going to have to do anything even that drastic—at least, right away. It's true that, given the way our economy is now set up, people need jobs. But these jobs are not the only jobs in the world, and this mountain is not the only source of gravel in the world. To the extent that more gravel is needed, it should be possible for quite some time to come to get it from places, and in ways, that don't involve conspicuously damaging highly visible natural or historic landmarks. (The mountain that started all this is both.)

The mere fact that a proposed project creates jobs does *not* automatically make it a good idea. There's money in bank robbery, too, but few would recommend it as a career for an ambitious young person.

In the case I'm alluding to, the faction trying to save the mountain had demonstrably thought about both economic and other considerations. They pointed out repeatedly and persuasively that, in the long run, this area is likely to prosper more from tourism and its spinoffs than from grinding up its mountains and selling them by the ton. The mine workers, on the other hand, showed no sign of having considered any facts except that mining the mountain would give them jobs that they might not have otherwise.

And that gravel would still be needed and would have to come from *somewhere*. Fair enough, for now; but I repeat, it does not have to come from one of the most conspicuous and scenic vistas, or one of the largest relatively undisturbed tracts, in the area. If this mountain isn't mined, the miners have at least two options. If mining gravel is the only thing they can imagine themselves doing, they can go where it's being done. Or they can learn to do something else.

Sure, those courses of action are not as easy or convenient as they might like. If alternate quarry sites are some distance away, working there may require a long commute or relocation. Changing to a different kind of work may require learning new skills. So what? Plenty of other people have had to do those things, and have survived the experience. Why should gravel miners be guaranteed a lifetime exemption from that possibility?

You can't ignore the guys with the pickets. They may be thinking only of economics, and some of us may consider other considerations more important—but it's a lot easier to feel that way if your economic necessities are taken care of. The miners are quite naturally and properly concerned about theirs, and it's unsurprising, if shortsighted, that they'd prefer to keep them taken care of in the simplest, most familiar way. If you know a reason why they can't or shouldn't, and alternatives they can pursue, you may have to tell them—in their own terms. Even if you think non-economic considerations justify doing something a certain way, it's prudent to be able to also couch your arguments in economic terms for those who don't understand anything else.

However, in the longer run, some more fundamental rethinking *will* be necessary—for all of us. The whole way of thinking that says you must keep indefinitely finding new sources of either jobs or gravel is invalid and must be changed. If you're *really* going to keep needing new gravel forever and ever, and Earth is the only place you can get it, you may be able to spare this mountain this year, but sooner or later you *will* have to mine it. If you pursue the miners' arguments to their logical conclusion, it's only a matter of time till *everything* has been mined for gravel, jobs, or similar commodities.

There are, of course, several options which can provide a better future than letting the miners and people like them do whatever jobs somebody offers them, regardless of non-economic costs. One that's probably obvious to readers of this magazine, at least in the area of gravel and other minable commodities, is space. There are *lots* of gravel in the asteroids, for instance. If you're going to keep needing more and more raw material, and you don't want to strip-mine every inch of Earth, sooner or later you're going to *have* to start getting some of it elsewhere.

Another thing we can do is learn to *use less* raw material. One way to do that is by recycling. If you need gravel to repair old roads, for instance, you may be able to get at least part of it by reusing what's already there, or what you get from old roads that have been abandoned.

Of course, if you want gravel for building *new* roads, you may have to get it from new sources. But you can't keep building new roads indefinitely, either. Eventually you'll run out of places to put them, and destinations worth driving to. Nor can we keep building more and more houses and business places indefinitely, though so far many of us seem determined to try. And the perceived need to do so will persist as long as people remain locked into the idea that they should keep increasing their own numbers without limit. They can do it for some time beyond where we are, of course; but the more of us have to share everything—material and otherwise—the lower the *quality* of life will become for all.

So we really need to break out of that mentality, that way of thinking that says we must keep making more and more people, and more and more roads and buildings to support them. And, finally, we must break free of the mentality that says that jobs are an *intrinsic* good rather than a means to an end. Yes, everybody needs a job, now—but the nature and extent of that need is a consequence of the way our society has evolved, not an innate law of nature. Originally people worked to meet needs—their own and their culture's. A socioeconomic system based on jobs for pay evolved to insure that those needs were met and that everybody (a) contributed to meeting those of society, and (b) earned the satisfaction of his or her own needs.

Now our society has outgrown the existing system. Its needs have changed dramatically, but it hasn't yet realized that fact or figured out what to do about it. Technology has greatly reduced the amount of personal labor necessary to meet basic needs, but the culture is still put together in a way that requires everybody to be employed for something approximating "full time." So a great deal of work now has nothing to do with basic needs, but is created primarily to provide employment. That work still consumes resources, and so a destructive—and quite unnecessary—cycle perpetuates itself.

We are facing an unprecedented opportunity that most of us persist in viewing as a problem. If and when we as a people recognize that fact, and figure out how to seize the opportunity—to restructure our lives so that most of us can get the things we need *without* a lot of unnecessary work-for-work's sake and wasteful use of resources—both we and our world will be much better off.

Invisible Enemies, Intelligent Choices

Americans' reactions to "evil spirits," whether the term is used literally or to refer to other threats not seen directly, such as microorganisms and radiation, continue to be one of our more reliable sources of amusement—and, on reflection, concern.

Take, for example, last year, when a couple of batches of ground beef were recalled because they were found to be contaminated with the bacterium *Escherichia coli*, a potential cause of gastronintestinal disease. The incidents were widely publicized, and widely followed by what can most charitably be described as panic. Beef sales dipped sharply, not just at the clearly identified suppliers that had the problem, but generally. Some people swore they would never eat beef again (though following up on how many *kept* that resolve might be another source of amusement).

Hardly anybody seemed to have the slightest grasp of the fact that the problem here was *E. coli*, not beef, or the fact that neither is necessarily associated with the other. There's no reason why beef *has* to be contaminated with *E. coli*, if it's processed and packed properly; and there's no reason why *E. coli* needs beef to be introduced into the human system. It can just as well ride on any number of other foods. (Actually, *E. coli* is a perfectly normal, permanent resident of the human digestive system. The problem with contaminated food is that its passengers may be a strain to which a particular immune system isn't acclimated, or they may be so numerous that they overwhelm the body's usually effective defenses.) All that seems to be much too involved for John and Jane Q. Public. Give a little publicity to one contaminated batch of *something*, and they promptly jump onto the Bandwagon du Jour: mortal dread at that *something*, with little or no attempt to understand the *real* (albeit invisible) danger.

It got even more interesting when people who were expected to do something about the problem proposed fighting one invisible enemy with another invisible thing that is also widely (albeit fuzzily) seen as an enemy. The Food and Drug Administration approved the use of irradiation, like that already approved but not widely adopted for treating certain other foods, to kill bacteria like *E. coli* and salmonella in beef. "The use of irradiation to deal with an otherwise intractable health problem is not only acceptable, but desirable," quoth one public health official.

But he obviously faced an uphill battle in getting the public to accept that notion, and the article in which I read his quote was just *full* of gems. One meat market owner said the use of irradiation would only hurt the meat industry, which, in the words of the article, was "already suffering from poor quality meat and poor packaging that can lead to contamination."

Which, of course, are exactly the problems that irradiation is intended to solve, and has shown itself, in several lab studies, to be quite capable of solving. The virtue of irradiation is that it can sterilize meat *after it has been packaged,* so it will *stay* sterilized, and without any known adverse effects on the food itself or anyone eating it. But that meat marketer is probably right about the prospects. When it comes to the public and anything containing the syllables "radiation," facts and logic have nothing to do with the case. People would much rather keep subjecting themselves to (and grumbling about) an invisible enemy that they *know* makes people sick and occasionally kills them, than fight it with something equally invisible that has never killed *anybody* and has so far shown no evidence that it ever will.

One shopper interviewed at a supermarket was quoted as saying, "When I think of radiation, I think of cancer. What's better, to die from radiation or *E. coli*?" Never mind that radiation is sometimes used (successfully) to *treat* cancer, or that the right kind of radiation used appropriately on *food* isn't going to kill *anybody*—and that a quick trip to the library, or any number of web sites, could make that quite clear.

A health commissioner correctly pointed out that, "Irradiating food is totally different from irradiating people. Irradiating food kills organisms that are harmful to people." But the reporter quoting him opined that the public would be hard to convince, adding, "After all, gamma rays are credited with turning comic book character Bruce Banner into the Hulk."

And in that sentence we see laid bare yet another invisible enemy that in the long run may do us far more harm than *either* microorganisms or radiation: galloping ignorance and sloppy thinking. The reporter didn't attribute that line to any particular individual, and my first reaction was to laugh at it. Then I realized that a great many people really *would* find that sentence about the Hulk at least as convincing as anything scientific "authorities" might tell them.

That is the invisible enemy I'm writing about here. *E. coli* and food irradiation are just recent examples of its *modus operandi*, not major issues in themselves. Personally, on the basis of what I've read, I'd like to have irradiated food available as an option; but promoting it is not by any means the main reason I'm writing this. I'm writing because there are so many things of comparable importance, and so many people who have vehement opinions about them despite having little or no understanding of them.

We may or may not need irradiation of food. We certainly and urgently need much better science education, in terms of both facts and attitudes.

Whether or not to accept and use any technology (of which food irradiation is just one example) is a *choice*, to be made by individuals and/or groups. As with any choice, I'd prefer that it be made individually whenever possible; there's simply no justification for my making a choice that affects only you, or vice versa. Irradiation is a case where the choice *can* be individual: If both irradiated and unirradiated foods are available, individuals can choose whichever they prefer, without any group imposing its tastes or prejudices on others.

There are, of course, other situations in which choices *cannot* be left entirely to individuals. Dumping untreated sewage into public water supplies, for example, does not have to be tolerated just because somebody feels like doing it, because its consequences clearly affect other people—who therefore have a legitimate say in whether it should be done. An interesting problem, of potentially life-and-death significance, can and does arise in such decisions: What if an ignorant majority believes it's perfectly all right to dump untreated sewage in the water supply, but an informed minority knows it isn't?

Whether decisions about a particular technology can be left to individuals, or must involve a larger group, the point is that there is a choice to be made. Therein lies one of the crucial distinctions between science and technology. We get a choice about technology, but *nobody* gets a choice about whether to obey natural law, which is indeed imposed upon us by a thoroughly inflexible Higher Power.

And choices about technology depend, among other things, on natural law (or, if you prefer, scientific principles). They *cannot* be made wisely without at least some understanding of those principles.

Please note carefully: I am *not* saying that all that matters about a technology is whether the hardware works, or that people should adjust to technology rather than technology to people. Of course technologies should be chosen and adapted to human needs—but a person with no understanding of how they work cannot realistically hope to do that. (Nor, of course, can a person with no understanding of human needs.)

Individuals and societies should, to the greatest degree possible, have the option of choosing the kinds and amounts of technology they want—but I'd like to be sure they are able to make those choices rationally. If the choice is purely individual, I won't even insist on that. If some people want to make their personal choices with no regard for facts or logic, far be it from me to deny them *that* choice. But even that should be an *informed* choice, and it can't if they know nothing of the facts or *how* to think logically. So, yes, I would like to see everybody's education include some solid grounding in principles of science and logic—so that even if they decide to reject it, they at least know what they're rejecting.

And if their choices affect not just themselves, but others—you and me, for example—then I *really* want those choices to be as informed as possible. If people want to make their own decisions on the basis of pure emotion, without any regard for whether they make sense on the basis of real principles or relationships, that's their business. But decisions that affect everybody are everybody's business, and the idea that anybody's opinion is as good as anybody else's just doesn't hold water.

Rights vs Rightness

Sometimes essayists, like dentists, hit nerves. Dentists usually aren't trying to; essayists often are—but sometimes we hit nerves we weren't even aiming for.

For example, in "Invisible Enemies, Intelligent Choices," I ended with the words, "Decisions that affect everybody are everybody's business, and the idea that anybody's opinion is as good as anybody else's just doesn't hold water." That drew letters from a number of readers who said they agreed with what I said but felt uncomfortable with the implications.

Well, good. That has long been one of the purposes of *Analog* editorials: to show up inconsistencies in people's beliefs by leading them down one path of reasoning to a conclusion that they want to reject. Ideally, the resulting discomfort should force them to resolve the conflict by thinking about which belief they *really* hold—or whether they need to reject both in favor of a third alternative.

In this case, I wasn't trying to say or imply anything about forms of government. I was simply stating a simple, irrefutable, almost self-evident fact: a right opinion (one that matches reality, such as "Fire burns") is worth more than a wrong one (one that doesn't match reality, such as "Fire has no effect on flesh"). But I'm glad some readers went beyond what I said to draw a disquieting conclusion about government.

That conclusion they drew was that, if all opinions are not equally valid, and good decisions must be based on valid opinions, then the ideal form of government must be a meritocracy, with decision-making limited to people who understand the issues. This disturbs people because it seems incompatible with the belief that all men and women are equal and should have an equal voice in government. As a particularly articulate reader named Beth Clarkson put it, "[the statement that all opinions are not equally good] is absolutely true. On the other hand, I have serious problems with anyone wanting to tinker with the idea of universal equality as the philosophical basis of our society. . . . America is founded on the . . . belief that 'all men are created equal.' This is a belief I hold quite dear despite its obvious falsity. I feel that a society in which people, and their opinions, are not considered equal is a society with a potential for abuse, injustice, and bad decision-making that is far worse than that incurred by the pretense that such a belief is true."

Well, maybe. Certainly there is potential for abuse and injustice in a society that considers some of its members' opinions worth more than others because of who holds them—particularly if it acts on that view by disenfranchising certain classes of people. Whether the dangers inherent in that abuse and injustice is worse than those inherent in decisions being made by ignorant masses is debatable and likely to vary from case to case.

And disenfranchising certain classes of people is not necessarily implicit in what I said, anyway. That inference depends on one or more hidden assumptions: that a meritocracy has to be small, and/or that the only way to keep ignorant voters from running the show is to disenfranchise them.

Quite likely that's enough hint that you can see where I'm heading, but please let me spend a little longer sneaking up on it.

We have two principles in apparent conflict:

1. **If decisions are going to significantly affect the lives of many people, it's important that they be made on the basis of a sound understanding and analysis of the factors involved.** In short, decision makers should understand what they're deciding about—which, ever more, means not only following the news and commentary in the media, but understanding something about how the physical

and natural world, and the technology we've derived from it, works.

2. **If decisions are going to significantly affect the lives of many people, it's important that all the people affected should have a voice in making them.** In short, everybody has a right to vote.

So which is more important: the right to vote, or the need to be right?

Ideally, many of us might say that both are highly important. In practice, we might recognize that any crowd of voters will include some who *don't* know enough to make a well-informed decision. So do we have to choose between entrusting our most important decisions to a small "elite" and entrusting them to a mob that, as a group, may have no idea how to make a rational decision?

At first glance, it would seem that we do. Hence the discomfort of my correspondents. Decisions like whether to dump raw sewage into our rivers (or, in a representative government, to elect people who are likely to do so) really *can* do tremendous damage to everybody, so it *is* important that they be made sensibly. But if we try to round up all our citizens who understand chemistry and ecology very well, and have them decide for us—the "meritocracy" my correspondents fear—we run just as great a risk of their abusing their power because the rest of us don't know enough to keep tabs on them.

Both dangers are serious. It's not clear that either is categorically more dangerous than the other.

So is there a way we can guard against *both*?

I think there is. The fact that my correspondents didn't mention it suggests to me that too many of us have given up on public education—and that's something we don't dare do.

Because general education that *works* is the one and only way I can see to head off both dangers at once.

I recognized the fundamental dilemma in "Invisible Enemies, Intelligent Choices," when I wrote, "An interesting problem, of potentially life-and-death significance, can and does arise in such decisions: what if an ignorant majority believes it's perfectly all right to dump untreated sewage in the water supply, but an informed minority knows it isn't?"

The answer is implicit in the question. If an ignorant majority is allowed to do something that will hurt everybody, then everybody is in trouble. If an informed minority is allowed to impose its desires on everybody else, everybody will quite likely soon be in a different kind of trouble.

Therefore our major decisions cannot safely be entrusted to *either* ignorant majorities or informed but self-interested and autonomous minorities.

Which leaves only one alternative: we *must* have an informed majority. A radical proposal, I realize, but the fundamental presumption of a successful democracy is an *informed, interested* electorate. Merely letting everybody vote, even if they know nothing about what they're voting on, provides absolutely no assurance of generally beneficial decisions. I seriously doubt that you would want your plumbing or dentistry done by a committee of people who know nothing about plumbing or dentistry. The folly of that is self-evident to most of us, yet we do very much the same thing when we let important decisions about the whole country be made without regard for whether the decision-makers know what they're doing.

If we want to stop doing that, we have two choices. We can abolish democracy and turn the decision-making over to panels of "experts," and hope that they're as interested in our welfare as in their own; or we can make a real, serious effort to make sure that most voters do know what they're doing.

There are at least two possible variations on that theme. One approach is to establish some means of requiring citizens to demonstrate minimal knowledge and competence as a prerequisite for the right to vote. This amounts to a meritocracy, as my correspondents feared, but it's a somewhat peculiar meritocracy: anybody can join it. All anyone has to do is take the trouble to learn enough to pass the test, whatever form it may take. There's no limit on size, and in principle this meritocracy could include the entire population.

In practice, of course, you could never get the experiment tried in today's social climate. (I won't say "never" in any broader sense, because all kinds of things have already happened in American society and politics that earlier generations would have found inconceivable.) Any sort of proof of competency would surely be compared to the literacy tests used in some of the darker episodes of our history to keep minorities from voting. The cries of "Racially motivated!" would be so strident that you couldn't get anybody to listen to the *real* motivation, or to suggestions for reducing or eliminating any ethnic side effects.

Which leaves us, in practice, with the *other* variation on How to Get Lots of Knowledgeable and Reasonable Voters: don't disenfranchise anybody, but increase the number of well-educated voters to the point where they can win elections by sheer force of numbers. That's not easy either, but it's what we need to do. It's a daunting prospect, when you consider the mess that constitutes so much of public education today, and the fact that the popular image of a well-educated person is inherently a caricature.

But consider the alternative: if we *don't* make a majority of the voting population take their job seriously, and learn enough to do it well, *you* are going to have life-and-death decisions about your future made by people who are incompetent to make those decisions. So *you*—no matter who you are—have a vested interest in making it happen. That will require improving schools throughout the land, even in such drastic ways as abolishing the concept of self-esteem as a birthright instead of something to be earned. I don't mean just rich people's schools or white people's schools or black people's schools; I mean *everybody*'s schools.

But it will take much more than that. It will also require changing *attitudes* throughout the land. We need to make being smart something for everyone to aspire to, not to ridicule. We need to show at least as much respect for somebody who can write an outstanding novel or explain something that's never been understood as for somebody who can hit a ball over a fence a lot of times in one season. The other side of the same coin is that willingly *not* learning should never be seen as "cool," but as stupid and pathetic. Social pressures in school need to be supportive, encouraging people to stay in and get all they can from the available resources. (Or get out and get all they can from the available resources. I'm well aware that some people learn better outside school than in it; the important thing is that we all need to learn, and keep learning, however we do it.) We must stop treating education as a low-stakes game, and start treating it—and voting—as serious business where *results matter.*

And none of us should be shy about pushing things in that direction. It *is* our business; every vote cast in ignorance hurts *us.* Changing such attitudes may seem a tall order, but it has been done in other areas. Look what's happened to the social status of smoking in the last couple of decades. If enough people want it to, it can happen to attitudes toward learning and citizenship, too. We *can* have both universal suffrage and intelligent government, but that's the only way I've thought of to get them.

And how many things can you think of that we need more?

Training Our Successors: Myths and Challenges of Education

Wishful Egalitarianism

It often seems to me that most people in this country are driven principally by fads—and educators, unfortunately, are no exception. Like other fads, those in education tend to fade out after a while, replaced by new ones which people hope will avoid the shortcomings of the old. Then they resurface a few years later, incorrectly relabelled "new."

One of the latest resurfacings is one that I'd really (if naïvely) hoped we'd seen the last of: "heterogeneous grouping," a movement to do away with "ability-grouping" or "tracking" in schools. Throwing everyone of the same age together in the same classes has been around before, of course, many times. From time to time it gets replaced by a system that attempts to assemble classes geared to students of similar abilities or achievement levels. The arguments for such ability-grouping include the claim that a group of very bright, highly motivated students can accomplish more if they are with others of similar abilities. They can pursue more advanced material faster if they don't have to wait for slower students who lack the ability or interest to keep up with them. Conversely, slower students can achieve more if they aren't intimidated or pressured to keep up with those who are much faster.

From extensive experience as both student and teacher, I consider these arguments to have considerable merit. But whenever ability-grouping has been in place for a while, complaints start being heard, gradually building up pressure to try something else. The something else, commonly touted as new but actually quite old, is "heterogeneous grouping," or putting pupils in classes without regard to their abilities or achievement levels.

Where I come from, we call this "ignoring relevant data," but the arguments advanced for it can be quite imaginative and occasionally amusing, in a grim, sad sort of way. One of the less ridiculous is the observation that ability-grouping, like type-casting, can become a self-fulfilling prophecy. Students placed in a group labelled "slow" may come to think of themselves that way, assume that it can't be changed, and play the role of slow learner for the rest of their lives. Those labelled "gifted" can, under some circumstances, get smug and figure they can rest on their laurels because they're "the smart ones."

It's a fact that these things can happen. It's also a fact that they don't have to—and that students really do differ in their abilities and achievements.

That's not a *popular* fact, in our present era of bizarre attitudes, but it *is* a fact. That's why the most fascinating argument that I've heard against ability-grouping is one that recently appeared in our local newspaper as a quote from the superintendent of a school system which is phasing it out. "If we don't do away with ability-grouping," he said, "we really can't have the belief that all children can learn."

Reread that closely. Evidently this person is more interested in propping up a belief than in giving people the best education. Sometimes, to some of us, *learning* as much as possible is more important than preserving a cherished belief—especially if the belief is wrong.

Oh, it's not completely and absolutely wrong; but it is wrong enough to do a lot of damage. From my own experience in teaching, I would agree that most people can learn, and in fact can learn more than they or most others think they can. But they can't all learn *equally*; and even if they could, they wouldn't—because some are more interested than others, despite the best efforts of even the best teachers. When I was teaching in classrooms, I considered it both an obligation and a point of pride to be able to do *something* for students all across the ability spectrum—but I could never deny that an ability spectrum exists.

It's not a one-dimensional spectrum, of course. People do have different strengths and weaknesses. I remember two physics students as a particularly good study in contrasts. One was extremely good at picturing conceptually what was going on physically, but unable to trust mathematical manipulations when the conceptual significance of every step wasn't easy to visualize (as one must often do in advanced physics). The other could manipulate symbols fluently and get any required mathematical result, but showed little conceptual understanding of what either the setup or the end result *meant*. Both did well, but in different areas and for different reasons.

In teaching several hundred students, I never met two whose minds worked quite the same way. For such reasons, there is probably a certain amount of validity in the desire of some teachers to have heterogeneous grouping so they can make up work groups consisting of students with complementary strengths.

But that validity is limited. The two students I mentioned were both intelligent people with the potential to be competent scientists, but each had strengths in specific areas that the other lacked. An "ideal" physicist would have all those strengths, but there are few ideal physicists (or ideal anything else). In a real work situation, either of those people might wind up on a team project, and a team with the two of them would function better than a team with two people like either of them. But it's hard to imagine a situation in which a team would be strengthened by pairing either of them with somebody who knew nothing about physics, had no aptitude for learning it, and no interest in trying.

And whether you like to admit it or not, there *are* such people.

It's probably true that a wide range of students can derive *some* benefit from heterogeneous classes—but not necessarily the benefits they should be deriving. The first function of a math class is to help people learn math, not merely to feel good about themselves and each other. It's become fashionable to put a great deal of emphasis on the importance of building self-esteem. Unfortunately, an essential ingredient often gets left out of that goal. If you just want people to feel good about themselves, you can do that by force-feeding them suitable drugs—but I don't know many people who would seriously consider that good, either for the person being euphorized or for the civilization in which he or she lives.

What's important is not simply to help people feel good about themselves, but to help them *deserve* to feel good about themselves. I found that the very best thing I as a teacher could do for a student was to help, cajole, trick, or do whatever else it took to get him or her to accomplish something he didn't think he could. It works wonders, not only for self-esteem (and perhaps we should try to ease the term "self-*respect*" back into greater currency), but for instilling confidence and desire to go on and accomplish still more.

But that requires that, for example, a math class must teach people to do *math*, not just fill them with a warm fuzzy glow. And everything I've seen compels me to believe that that is far more likely to happen in ability-grouped classes than in sociable hodgepodges. Just what, you ask, have I seen that would make me say such a thing? Well, before teaching a variety of classes by a variety of methods, I spent a lot of time as a *student* in classes at all levels from kindergarten through doctoral.

In particular, I was in junior high and high school at a time when ability-grouping was just coming into one of its several vogues. As a result, my school offered some "accelerated" and "advanced placement" courses, but only in a few selected subjects. So I had some of those, and some regular, heterogeneous classes where students were simply scheduled where they would fit. I'm in an excellent position to make a side-by-side comparison of the two kinds of classes, from the viewpoint of a student who could and wanted to learn.

My verdict? The "advanced" classes made high school endurable; most of the "mixed" classes required all the endurance I could muster. The advanced classes actually taught me things I hadn't known. The mixed classes, in most cases, taught me little, if anything. And the difference wasn't just in the curriculum or the teachers, but in the company. The "regular" classes had so many "students" who were more interested in disrupting the class than in learning anything that the few who *wanted* to learn had an uphill battle. The advanced classes, on the other hand, were the first places where I had ever been able to sit in a whole room full of people who wanted to learn, respected learning, and *could* learn. That was an exhilarating, stimulating experience, and worth the price of admission even when (as occasionally happened) the formal curriculum content seemed silly or counterproductive.

So I must say, no, Mr. Superintendent, you *can't* have the belief that all children can learn, at least equally; and I'm appalled to hear that you are yet again determined to inflict on real children a system based on a wishful premise. All people are *not* created equal, except in certain very special senses. Effective education has to work with their real differences, not pretend they don't exist.

What about that concern that people once launched on one track will be stuck there forever, coasting or falling ever farther behind? It's a real concern, but it does not justify being afraid to build different tracks. It simply means that when you're doing so, you must also build in a mechanism for switching from one track to another. And you must make sure the teachers on *all* tracks watch their students closely enough to know when a track-jump should be considered. Students on the "fast track" must know that it does not mean a free ride from here on out; they must continue to perform on that level or they'll have to get off. And students on the "slow track" must know that they are not stuck there irrevocably. They must know that it is possible to move to a faster one if they show the ability and interest, and their teachers must watch for that ability and interest and nurture it wherever they find it.

And if all these things are done conscientiously, I'll bet on the resulting system to give a better education to more students than any system built on the wishful thinking that all people are equal in ways that they aren't.

I/O

In August of 1992, *Analog* published Michael F. Flynn's haunting story "Captive Dreams," about a young boy who was obviously retarded—but in a far more literal way than anyone first suspected. The thinking part of his brain was as good as anybody's; his problem was that his "input/output buffer kept getting backed up," so that most of his sensory inputs reached his brain seconds after the stimuli that produced them. Thus his responses were usually inappropriate, and the feedback he got from them merely added to his confusion. He would, for example, reach for a whirling pinwheel he'd seen, but his hand would close on the empty air from which it had since been moved.

It's not enough to know information, or to know how to process it. To *use* it, you also have to be able to get it to and from the outside world by appropriate channels and on an appropriate time scale.

Flynn's story was by no means the first science fiction to deal with ideas about input and output and their importance to the human nervous system. Many stories have dealt with things like direct brain-computer interfaces: implants that give everyone instant access to the equivalent of huge libraries, for example, or war machines whose control centers are built-in human brains.

Nor are such concerns confined to science fiction. In the real world as we have already experienced it, many stroke victims have experienced the frustration of knowing what they want to say or do and being unable to act on the desire or communicate it to those around them. Those who have lost one sense, such as sight or hearing, must learn to compensate by getting information through other channels more efficiently than most of us need to. Autism and dyslexia are disorders in the way a brain interfaces with the outside world.

In fact, concerns about input and output are confined to neither fiction nor pathology. Getting information to and from a person's central processor plays a key role in virtually every aspect of normal living.

Using language, for instance. I used to think I could learn a language from a book, and to a certain extent, I can. But even if I've learned enough from books to be able to read a newspaper or magazine rather fluently, the first times I attempt conversation—speaking and understanding real people speaking normally—are likely to be humbling experiences. Knowing *how* to speak a language is not the same as being *able* to speak it.

Apparently what happens is something like this: If you study a language entirely by reading books and writing exercises, you establish neural circuitry that directly associates written words brought in by your eyes, or sent out by your hands, with meanings and grammatical structures in your brain. You may know how to pronounce all the words, but if you haven't similarly developed direct neural linkages between your brain and your mouth or ears, you'll have to translate everything you hear into its written form before you can understand it.* Real conversation doesn't allow time for that, so conversation can be hard even if reading and writing are easy. Similarly, the "total immersion" approach to language teaching trains *only* the conversational linkages, and doesn't guarantee proficiency in reading or writing. Probably the best way to learn a language is one that uses *all* available channels. That not only provides the whole set of useful skills, but likely speeds the firming up of the "CPU" circuits that

*Please note that I *didn't* say you must translate everything into your *native* language, as too many poorly taught language courses allow or even encourage you to do. To learn a language well, you must train groups of nerves to directly associate symbols in that language with their meanings. But the nerves that associate *written* symbols with their meanings are not the same ones that associate *spoken* symbols with their meanings!

think in the language, by letting the various channels reinforce each other through cross-linkages.

Other examples of the importance of neural input and output mechanisms abound. In music, knowing how to play an instrument is not the same as being able to play it well. A high school band director must know how to play all instruments in his group—he must be able to answer a clarinetist who doesn't remember a fingering or a trombonist a slide position—but may actually be proficient on only a small number of them. Knowing fingerings and positions merely involves storing data, which, for question-answering purposes, can be accessed in a relatively slow and clunky fashion. *Playing* fluently requires a direct link from brain to mouth and fingers.*

As a writer, I not uncommonly find myself getting so close to a deadline that I *must* start writing a piece even though I don't feel ready to. That is, if I try to talk or think ("talk silently to myself") about what I'm going to say, I don't seem to have a clear enough idea of what that's going to be. If I start trying to put something on paper or disk anyway, because I have to, I'm often surprised at how easily the ideas flow and how well developed they are. Evidently the creative part of my mind (which functions largely at a subconscious level) has already done more work than I realized, but its "hot line" to my keyboard fingers is much better developed than its connection to my other output devices. (I suspect this implies that I would find writing by talking into a tape recorder, as some writers prefer, awkward and frustrating—until I trained a new output system.)

Speaking of keyboards, I sometimes find myself called upon to explain to someone else how to do something unfamiliar to them on a computer. Even if the operation in question is something I can do

---

*Some players must even train several *different* sets of linkages. For complicated historical reasons, orchestral trumpet and horn parts are often written in a different key than the one in which they must be played—and the way in which they differ is not always the same. Thus trumpeters and hornists must learn to transpose in several different ways, and a player who has to consciously *think* about how to transpose each note will not be able to keep up.

quite easily, I'm always amazed at how much harder it is to talk someone else through it than to sit down at the keyboard and do it myself. Here again, I've evidently trained neural circuitry that directly translates thoughts into finger motions. Translating them instead into verbal descriptions of those finger motions is an extra and, it would seem, a surprisingly roundabout and cumbersome process.

Slower, of course, is not always, necessarily, or intrinsically *worse*. Sometimes speed is important and sometimes it isn't. I'm not sure whether it was really Albert Einstein who first said, "Never learn what you can look up," but I wouldn't put it past him. Whether he did or not, certainly many others have—including me, on occasion. It's quite appropriate for some circumstances, but not all. There's not much point in cluttering your personal memory with lots of detailed facts that you'll need only occasionally, if ever. For most of us, for example, it would make no sense to memorize the density of water to six decimal places at every temperature between freezing and boiling. If you ever need to know it at a particular temperature, you can look it up; and you probably *won't* ever need to know it at most temperatures. If you're working on a refinement of theory for which the detailed way the density varies with temperature is crucial, you will likely become very familiar with the shape of the curve, and you will at some time need to know a goodly number of precise values. But you probably *won't* have to carry them all around at your mental fingertips. New scientific theories are rarely needed or expected on a moment's notice, and good ones are worth waiting for.

On the other hand, some things *do* need to be at your mental fingertips. To speak a language fluently, you *cannot* take time to look up every word and grammatical rule you need to construct a sentence like, "Jump away from there before that safe lands!" To play fast passages on an oboe or violin, not only can you not take time to look up the fingering of each note, you can't even take time to think about what each note *is*. Not only must the meaning of each note on the printed staff be at your mental fingertips, but your mental fingertips must be connected as directly as possible to your physical fingertips.

To fly safely from New York to Los Angeles, you would not want to depend on a pilot who has to look up the procedure for increasing his plane's airspeed, or what the airspeed should be for a particular maneuver.

All of which implies, by the way, that if you want to test how well someone has learned something, a truly meaningful test must require them to output the data in a way as close as possible to the one they'll use in a "real-life" situation. A written multiple-choice test on bassoon fingerings or how to land an airplane may demontrate presence or absence of knowledge needed for those tasks, but it *cannot* prove *competence* in them. The only way to do that is to require the ~~victim~~ student to play a challenging bassoon solo or land an airplane safely.

Come to think of it, how many "real-life" situations can you think of in which information *is* accessed and outputted in any way resembling a multiple-choice test?

Maybe it's time for people who really want to improve education to be paying a lot more attention to the detailed neurophysiology of input and output systems, not only for designing more meaningful tests, but for finding more effective ways to get information stored durably and accessibly. Much discussion of education concerns *what* should be taught. We also hear lots of advocacy of this or that teaching method, some of which catch on enough to become fads for a while; but in my experience, most of the "revolutionary improved methods" are based far more on somebody's hunch or wishful thinking than on any actual knowledge of the mechanisms involved. It may be that to effect much improvement in education, more attention needs to be given to the nitty-gritty of how knowledge gets in, and how it will have to come back out.

Style and Substance, Horse and Cart

Some years ago ("Speak for Yourself," *Analog*, April 1985), I made the heretical suggestion that political campaigns should be conducted entirely by means of candidates explaining their positions *themselves*, in their own words, with none of the showmanship now generated for them by professional publicists, ad agencies, and speechwriters. This was a response to the observation that elections these days are based less on what candidates think and can do than on how flashy their ad campaigns are.

It's still an idea worth considering, I think, but it probably doesn't go far enough. Even if a condidate *does* speak for himself (and I use "himself" throughout this discussion in the well-established sense of "himself or herself"), the content of his speech is likely to have far less impact than its delivery. And there are professionals willing and eager to coach him on *that*, too.

I was pointedly reminded of this while flying home from the 1996 World Science Fiction Convention. The inflight magazine in the seat pocket in front of me contained an article by Marion Winik called "Welcome to the Sound-Bite Factory," about a "media training" business called "On Camera." And that contained this chilling tidbit: "We're all such sophisticated consumers of communication, inundated with molded messages and sculpted images, that it's really not okay just to 'be yourself' if you're in the hot seat. . . . Studies show that ninety-three percent of what a listener gets out of any communication has to do with the demeanor, appearance, and likability of the messenger . . . ; only seven percent is the message itself."

Hence a booming business in "sound-bite factories," schools to teach politicians, corporate executives, touring authors, and so forth to glitz up their demeanor and appearance and project likability on camera and microphone. The article I read describes in considerable detail how one such school works, by describing the first-person experience of one individual who's been through the program and quoting others.

The experiences described are quite consistent with the underlying thesis that presentation is almost everything and content virtually nothing. (Abraham Lincoln wouldn't have had a chance in today's political marketplace!) The procedures described put all their emphasis on style; *nothing* is said about content.

Unfortunately, that 93 to 7 division of emphasis may indeed be an accurate description of how the American public evaluates what it sees and hears. If so, the "sound-bite factories" may indeed be effective in helping their clients to win elections, promotions, and impressive sales. Certainly they're a nifty way for the people running them to make lots of money.

But are they good for the *country*, or the larger civilization the country is part of? Or even the company or individual paying for the image-building service?

In the short term, and on a small enough scale, sure, they work. An author doing a book tour has no cause for complaint if taking one of these courses increases his royalties by more than the cost of the course. A politician has no cause for complaint if it helps him get elected. A business executive will surely rejoice if a slick presentation helps him get the board to approve his pet project instead of somebody else's.

But how do that author's *readers* fare if they buy the book on the basis of the hype and then find that it's a lousy book? What if the politician so elected turns out to be incompetent and/or crooked, and undermines a multitude of good works started by the less flashy incumbent he unseated? What if the executive's project is unsound and loses his company a lot of money, while the one it beat out could have made it a lot of money?

I think the answers to those questions are fairly obvious. What may be less obvious is that, in the longer run, even the individuals who triumphed through professionally cultivated glitz may suffer. There is, after all, that seven percent of judgment based on content. If the author's book is bad enough, readers who were stung once may be harder to take in again—even if his next book is better. If the politician does enough damage, eventually enough voters might notice to turn him out. The executive may find it harder to sell his next scheme.

Of course, 93 to 7 is pretty good odds for style over substance. So it may take quite a while for any of those things to happen. Thus, on balance, the alleged fact that people judge what they see and hear far more on presentation than on content, if true, represents a very serious problem. People who understand it and concentrate hard on presentation have an excellent chance of using it to get themselves into powerful positions, and then stay there long enough to do lots of damage.

It's been often (and rightly) said that one person's problem is another person's opportunity. If most people judge what they hear almost entirely on the basis of its presentation, that's a problem for *everybody*. It implies that we-as-a-culture are going to buy a lot of things that are bad for us, simply because they had slicker sales pitches.

But "everybody" is hard to reach. The style-to-substance judgment ratio is also a problem for individuals with something—whether a "how-to" book or a foreign policy—to sell: A "non-grabbing" style will make the sale more difficult. But such individuals are much easier to reach and deal with—and so the problem viewed this way is much easier to turn into an opportunity. If presentation is polished enough, content hardly matters. Individuals who recognize that fact can turn it to their advantage by learning to polish their presentation, and "sound-bite schools" can turn it to *their* advantage by helping them do so.

Which is all very well, at least in the short term, for Those Who Would Sell Something. The fact remains, though, that the far more important concern in the bigger picture is: Is what they're selling worth buying? Readers want books that they actually enjoy reading. Companies need products that actually satisfy customers and keep them coming back. Countries need policies and leadership that actually contribute to the well-being of their citizens.

And whether they get those things depends far more on content and substance than on style and presentation.

So I view the "sound-bite factory" phenomenon as putting its emphasis in a perversely wrong place—or, at best, as a very incomplete solution to the larger problem. Given the apparent reality that most people do judge far more on style than on substance, even people with excellent ideas and products may need the skills they can learn from such programs to get them accepted. To the extent that they can help that happen, I can (grudgingly) acknowledge and appreciate their value. At the same time, I can be very disturbed by the role they can play in helping people sell useless or destructive products, ideas, and policies.

The much bigger, and far more important, challenge is to make *that* less likely—and that won't be solved by expensive schools teaching a few well-heeled clients to make slicker and slicker sales pitches. The real challenge is to *change* that ratio: to train a far larger part of the populace to judge what they hear *not* almost exclusively by how it's said and how the person saying it looks, but primarily by what it actually *means* and how believable it is on the basis of its own merits. The justification for "sound-bite factories" I quoted earlier starts out by saying, "We're all such sophisticated consumers of communication . . . ," but I think the real situation is just the opposite. If we're so easily taken in by carefully cultivated posturing and mannerisms that content counts for almost nothing, we are terribly *un*sophisticated consumers of communication, and as such terribly vulnerable.

Let the sound-bite schools continue; as long as this situation persists, even those with something good to sell may need them. But what we need far more is *other* schools that will do everything in their power to immunize the general populace to the slick tricks of the "style over substance" crowd. We need schools that will teach *everyone* to see through the tricks, to cut past facades and think critically about what people are actually *saying*—and to judge them primarily by that, not by how they look or how glibly they speak.

The task of *those* schools will be, of course, far harder. They need to reach far more students—ideally, they should include every public and private school for general education—and many of those students will be far less motivated than those at the sound-bite factories. The work will be harder (any ideas on *how* to do it are eagerly solicited!) and financially less rewarding.

But I can't think of many jobs that more urgently *need* to be done.

Relevance

Back around 1970, there was a sudden burst of demands by students for "relevance" in their studies. They didn't want to be bothered with subjects that didn't have a direct and obvious application to the social problems of the day, such as the Vietnam War and the suddenly popular concern for the environment. A lot of people were *very* concerned about such things then; and many teachers, seeing no realistic alternative, acquiesced at least somewhat to the demands.

The demand for relevance was, in part, a fad, and, like so many others, eventually subsided. It has seldom been heard quite so widely or so stridently since—but it didn't go away completely. We still hear mutterings, from time to time, that this or that—space travel, for instance—is not worth doing because it's not "relevant." Relevance *is* sometimes important—but not necessarily in the senses and for the reasons that those students had in mind.

From a teacher's point of view, relevance is not so much a quality that everything *needs* to have, as one that can serve as a valuable motivational tool when something *does* have it. You must remember, first of all, that the concept is meaningless in a vacuum. Things are not *intrinsically and absolutely* relevant, but only relevant *to* something else. There are so many elements and relationships in the world that most things *are* relevant to at least something else, and many teachers need to make more effort to point those connections out. A student who is interested in *A* but not *B* may suddenly pay more attention to *B* if he realizes he can use it in *A*.

Even very good teachers sometimes miss, or forget, this point. I remember being perplexed as an undergraduate by the very different opinions my classmates held of a particular professor. Some of us thought he was outstanding. Many others thought he was hopelessly boring. How could we be talking about the same man?

I think I figured it out, much later. Those of us who admired him came to the class wanting to learn physics, and he had a knack for making very clear the things we wanted to know. But he made little effort to create an interest in the subject in students who didn't already have one, and those were the ones who found him boring. He was an extremely good teacher for those of us who knew we wanted to learn his subject—and, I must now sadly agree, not a very good one for those who didn't.

He could have been an even better teacher by reaching more students, but to do that he would have had to make an additional effort to "wake up" the ones who didn't want to be there. One way he could have done that would have been to explicitly point out ways that his general principles applied to other fields that did interest them—how, for example, principles of hydraulics directly govern the circulatory systems that would occupy so much of the future professional attention of premeds.

There is certainly room for debate about how much of a teacher's attention should go into trying to coax and cajole students who have failed to figure out why they're there. My personal view is that such efforts should not be allowed to dilute the substance of what's delivered to those who want to learn. On the other hand, the empirical fact is that some classes (such as physics for premeds) routinely contain a high percentage of "captive" students. When I'm teaching such a class, I can't believe I'm doing a very good job if I use their reluctance and ignorance as an excuse to let them out without learning anything. So when I taught such classes (and I sometimes *asked* for them because I considered them a challenge), I tried to provoke both interest and accomplishment in as many students as I could. I used whatever methods I could find that worked, and demonstrating relevance to their own strong interests was one of the most reliable.

I can also think of many examples from a student's point of view—both things that *were* done, and things that I later *wished* had been done. In the early stages of learning to fly an airplane, I sometimes wondered why so much of my time was spent being made to do difficult things bearing little resemblance to things done in the normal course of flying. What was the point, for instance, of spending a whole hour of instructional time making the plane stall in various ways, or flying tight circles around a barn on a windy day? Eventually I saw that all those weird exercises were really aimed at developing skills needed for the one supremely important and difficult thing a pilot must do on every trip: land the plane in one piece. Might I have found those early hours less trying if somebody had *explained* that to me then?

Maybe. But the main value of such advice is as a tip to teachers on how to keep students more interested through distasteful things they must learn, by telling them *why* they must learn them. Where the students demanding relevance sometimes went too far—and so did some teachers who let themselves be pressured—was in believing that students should unilaterally decide what is relevant and worth studying. Letting them do that is unrealistic, and likely to cheat the students themselves in ways that even they would recognize eventually—but too late. Students, by the very nature of the incompleteness of their present knowledge, are not always in a good position to *know* what is relevant or important.

The flying example is but one illustration. A student who hasn't been told why practicing stall recoveries and circles around a point is important might well grumble that they're "irrelevant" and demand to skip them and get on to the fun stuff. But an instructor would be a fool to listen to such demands. A landing is a precision maneuver done at or near stall conditions, and it *must* be done *right* under all kinds of wind conditions. So every pilot must be very familiar with what stall conditions feel like and how to cope with varying winds. Those "irrelevant" maneuvers provide excellent ways to learn those things at a safe distance from the ground.

It simply isn't possible to judge how you're going to like a field when you haven't tried it, or apply a technique when you haven't yet learned it. Differential equations have never been popular, even among physics students, but you can't go very far in physics without *needing* them on an everyday basis. My wife now wishes somebody had made her study more music, and some Polish (her ancestral language), when she was little; and I now agree that it would have been to her present benefit if they had.

In my own case, I early developed an interest in musical composition, with fairly grandiose ambitions: I wanted to write symphonies. The kinds of symphonies I thought I'd like to write were more like those of Shostakovich and Mahler than those of Mozart and Haydn—but years later, I found myself wishing somebody had made me study the structure of Mozart and Haydn's whether I found them boring or not. The *basics* of form are much easier to see in the earlier composers, and the later ones are easier to understand if you can see how their methods *evolved* from the earlier. But of course I couldn't see that until I'd learned quite a bit about both, by a harder method than necessary. . . .

What it boils down to, I think, is this. In education (as in many other areas) a double standard—in the right sense—can actually benefit everybody. Students should make sure their teachers know as much as possible about what they're interested in, but they should recognize that their teachers may know something they don't about what's needed to pursue those interests. Teachers should bear in mind that it's their obligation to provide what students are going to need, whether they currently recognize the need or not—but part of the job is to show them *why* it's a need.

Finally, everybody—whether or not they're currently involved with formal education in *any* capacity—should bear in mind that you can't always *know* what will turn out to be important sometime in the future. Somebody once asked one of the pioneers in the development of electricity, "What good is it?" His reply was, "What good is a newborn baby?"

Personally, I've always been an informational pack rat (which, as it turns out, is a very useful thing for an *Analog* editor to be!). I tend to save more information than many people do (or think reasonable), and over and over I've found some bit of it proving useful many years later—not for what I'd anticipated it *might* be useful for, but for something I couldn't have anticipated at all. I saved notes from astronomy courses I taught years ago, figuring I might use them as a starting point if I ever found myself teaching such a course again. So far I haven't; and if I ever do, the new course will need a lot of updating—but the old notes were surprisingly helpful in thinking about the organization of a book I unexpectedly found myself writing last year. Over the years I've given many slide shows, on a variety of topics for a variety of audiences, and usually saved scripts in case I ever found myself called upon to give the same show again. That's very seldom happened—but those old scripts much later turned out to be a great labor-saver in developing a computer aid for editing *new* shows on new topics.

If you *know* something is relevant to something you want or need to do, that gives you an extra incentive to pay attention to it. If you *don't* see its relevance, that proves only that you don't see it—not that it doesn't and never will have any. In a rapidly changing world, making "relevance" one of your primary goals is a dangerously short-sighted way to live. If you insist that everything you do or learn be relevant to the problems of today, you may leave yourself utterly unprepared to deal with those of tomorrow.